Building on the Kyoto Protocol:

Options for Protecting the Climate

Edited by

Kevin A. Baumert

With

Odile Blanchard
Silvia Llosa
James F. Perkaus

World Resources Institute

October 2002

CAROL ROSEN
PUBLICATIONS DIRECTOR

HYACINTH BILLINGS
PRODUCTION MANAGER

MAGGIE POWELL
LAYOUT

KAREN HOLMES
EDITING

Each World Resources Institute report represents a timely, scholarly treatment of a subject of public concern. WRI takes responsibility for choosing the study topic and guaranteeing its authors and researchers freedom of inquiry. It also solicits and responds to the guidance of advisory panels and expert reviewers. Unless otherwise stated, however, all the interpretation and findings set forth in WRI publications are those of the authors.

ISBN: 1-56973-524-7

Library of Congress Control Number: 2002113906

Printed in the United States of America on chlorine-free paper with recycled content of 50%, 20% of which is post-consumer.

Photo Credit: Paul Souders/Getty Images

Contents

Foreword

Now that the 1997 Kyoto Protocol is poised to enter into force, what comes next? How can the international community address the growing threat of climate change in a way that is fair and effective?

In *Building on the Kyoto Protocol: Options for Protecting the Climate*, Kevin Baumert and his colleagues analyze a thought-provoking spectrum of possibilities for shaping an international climate change agreement. Seventeen contributors from nine countries offer analyses of options for strengthening the climate protection treaties. They confront the most persistent challenge of climate protection—designing solutions that include both developed and developing countries. The options examined range from the well-known to the novel. Indeed, some approaches put forth in this book have never before been examined in print.

This is a good sign. We need innovation and cooperation if we are to create environmentally sound solutions that are economically and politically viable. From the vantage point of the 1992 Earth Summit in Rio de Janeiro, a climate treaty with simple timetables for emission reductions may have seemed plausible. Some thought a global carbon tax would provide a simple answer. Today, we have a better sense of the limited capacity of our global institutions and national governments to deal with a problem on the scale of climate change. The linkages between economic development and climate change are complex and resist simple fixes. Several of the approaches examined in the following pages tackle development and climate protection simultaneously. The authors use case studies—Mexico, South Africa, and South Korea, for example—to show how the concepts are applied in national contexts.

What comes after Kyoto? Citizens and governments should not be satisfied with a haphazard round of piecemeal commitments. The dangers of climate change are too great, and those dangers fall disproportionately on the poor. This book outlines a way forward: a two-track strategy that both meets the short-term interests of governments and sets in motion processes for developing a more coherent long-term framework for climate protection. The final chapter distills key lessons that government negotiators and civil society advocates can use for many years to come as the international community debates how to protect the climate system from dangerous human interference.

The origins of this book also help illuminate its content. It grew out of a unique and compelling collaboration, called the Climate of Trust, founded in 1998 under the leadership of Bonizella Biagini, Atiq Rahman, Nicolás di Sbroiavacca, Agus Sari, and Youba Sokona. Like the current Climate of Trust collaborators convened by WRI, the original partnership firmly believed that it was possible to establish an atmosphere of trust between developed and developing nations, especially with respect to the urgency of addressing climate change. The first Climate of Trust report, *Confronting Climate Change: Economic Priorities and Climate Protection in Developing Nations*, was published in 2000 by the National Environmental Trust and Pelangi. *Building on the Kyoto Protocol* is the next step, and includes some new research partners from both developed and developing countries, to expand the circle of individuals and organizations committed to creating solutions that span the North-South divide.

Most of the authors contributing to this book are from developing country research institutes. While the industrialized countries must take the lead in reducing global greenhouse gas emissions, developing countries can and must play a leadership role in shaping durable and workable solutions to climate change at the international level. Indeed their future development prospects may depend on such solutions. By building bridges across the North-South divide for research and analysis, this project helps to reduce that real-world divide; to foster trust and spark a broader dialogue that can serve as the basis for concerted government action.

Support for this book and the Climate of Trust project comes from the Canadian International Development Agency, the Helen Brach Foundation, the Institut Français de l'Energie, the Italian Ministry of Environment and Territory, the John D. and Catherine T. MacArthur Foundation, the Netherlands Ministry of Foreign Affairs, the Swiss Agency for Environment, Forests and Landscape, and the Wallace Global Fund. I am grateful for their generosity and foresight.

Jonathan Lash
President
World Resources Institute

Acknowledgments

This book is the product of a true collaboration. We are especially grateful to all of the *Climate of Trust* participants that authored the individual chapters of this book and participated in many aspects of this project. The team of authors—drawn from nine countries—participated in nearly every step of the peer review, revision, and editing process and adhered to deadlines that, at times, were tight. The authors also assisted one another by providing input during the peer review process and during a March 2002 authors' workshop in Rio das Pedras, Brazil (where preliminary papers were reviewed and discussed). The editorial team especially benefited from the authors' feedback on the Introduction and Conclusion chapters. It was a privilege to work on an ambitious project with such a dedicated and creative team. It is the actions of the project participants, more than their written words, that are helping to create an atmosphere of trust between developed and developing countries.

We owe particular gratitude to several people who helped guide and advise us on the overall project. Nancy Kete's core intellectual input provided this project with a firm foundation. Her skillful facilitation made the author's workshop a success. When challenges arose, Nancy was there to point us in the direction of relief. In addition to being an author, Christiana Figueres also provided important guidance and advice early in the project, helped moderate the authors' workshop, and reviewed early draft papers. We owe a special thanks to Bonizella Biagini—friend, former WRI colleague, and a founder of the *Climate of Trust* project. Boni inspired WRI to "build a climate of trust" through this project and was instrumental in the early development of our research and planning. The *Climate of Trust* was originally created by a partnership of non-governmental organizations from around the world in 1998, led by Boni Biagini, Atiq Rahman, Nicolás di Sbroiavacca, Agus Sari, and Youba Sokona. We are grateful for the example that the original *Climate of Trust* partners provided for us to follow.

The manuscript benefited enormously from peer review. We are thankful to the following reviewers for their helpful comments and suggestions: Tom

Athanasiou, Marcel Berk, Hadi Dowlatabadi, Michel den Elzen, Michael Grubb, Niklas Höhne, Cathleen Kelly, Karen Lawson, Rodolfo Medina, Aubrey Meyer, Benito Müller, Cedric Philibert, Michael Philips, Atiq Rahman, Liam Salter, Anju Sharma, Richard Sherman, Fernando Tudela, and Julio García Velasco. We are particularly grateful to Jan Corfee Morlot, Elliot Diringer (with Sophie Chou, Christie Jorgé, and Neil Strachan), Jennifer Morgan, Granger Morgan, and David Victor for investing the time to review all (or nearly all) chapters of the book. The many helpful comments and suggestions we received substantially improved the manuscript. Any remaining errors and omissions are, of course, the responsibility of the authors and editors.

Many of our colleagues at WRI helped make this book possible. Early versions of the manuscript were reviewed internally by Navroz Dubash, Christine Elias, Suzie Greenhalgh, Tony La Viña, Julia Philpott, and Chris Whaley. Barbara Haya provided continuous and thoughtful contributions to the Introduction, Conclusion, and several other chapters. Jennie Hommel assisted in almost all aspects of this project, including the planning and logistics for the authors' workshop, organizing the review process, overseeing contracts, and smoothing the publications process. Alex Ruthman compiled the comprehensive bibliography. Gracie Bermudez and Belle Schmidt made the external review an efficient and pleasant process. Carol Rosen, Bill LaRocque, and Hyacinth Billings advised and guided us during the publication process. Karen Holmes did a fantastic job editing the manuscript and improving the prose, and Maggie Powell is responsible for layout. Finally, Navroz Dubash more than capably served as the review steward for this book, helping us to navigate the internal and external review processes. His careful eye and sharp mind improved the final product. We are grateful to all of our WRI colleagues and the organizational support we received.

Finally, without financial support, this *Climate of Trust* collaboration and the resulting book would not have been possible. We are especially thankful to the Canadian International Development Agency for their financial support that launched this project. The project also received generous support from the Helen Brach Foundation, the Institut Français de l'Energie, the Italian Ministry of Environment and Territory, the John D. and Catherine T. MacArthur Foundation, the Netherlands Ministry of Foreign Affairs, the Swiss Agency for Environment, Forests and Landscape, and the Wallace Global Fund.

Kevin A. Baumert
Odile Blanchard
Silvia Llosa
James F. Perkaus

1. INTRODUCTION: *An Architecture for Climate Protection*

Kevin A. Baumert and Nancy Kete

Walls, windows, floors, and doors are some of the elements used in designing a house. Not just any mix of architectural elements will create a functional home. Although there are many possible designs, making a home functional means assembling walls, windows, and other elements into a compatible whole that meets the needs of its inhabitants. So too, a climate protection treaty has its own set of "architectural elements" that must meet the needs of its stakeholders. Elements of a climate protection treaty include provisions for controlling greenhouse gas emissions, managing economic costs, and promoting accountability, among other things. While the options for designing a home may seem limitless, the diversity of potentially effective climate agreements is not nearly as constrained as current international negotiations might lead us to believe. As with homes, innovation and creativity are needed in treaty design.

Since 1997, the debate over global climate change has focused narrowly on the Kyoto Protocol—an international treaty to control greenhouse gas emissions that are trapping heat in the Earth's atmosphere. The Protocol calls on industrialized countries to reduce their emissions of greenhouse gases by about 5 percent below 1990 levels between 2008 and 2012. Over the past 5 years, government officials, observers, and experts have been absorbed in the arcane details of the Protocol, arguing at great length about the treaty's merits and demerits. As the Kyoto Protocol comes to life, this debate will shift to include new ideas for future commitments to protect the global climate system.

This volume explores a set of options for designing an international framework for climate protection "beyond Kyoto," that is, beyond the Kyoto Protocol's first commitment period.[1] We pay special attention to achieving international cooperation across the so-called North-South divide. Each approach examined in this volume could embrace both industrialized and developing countries—an eventual necessity for addressing the problem of climate change. In conducting this study, we hope to promote a better understanding of a wide range of future climate protection options and provide building blocks for further consideration of these ideas and other alternatives not yet considered. An improved understanding of future options, we think, will lead to more environmentally effective and fair outcomes under the international climate change negotiations.

We believe now is a particularly opportune time to think expansively, creatively, and critically about different approaches to protecting the global climate system. With every year that passes, we are wagering the future, betting (or simply hoping) that global warming will not manifest itself through the worst possible outcomes before humanity finds a collective and effective long-term solution. We need intellectual and creative resources from across the globe to produce and test ideas. Just as shooting more arrows at a target provides a better sense of how to hit the bull's-eye, the more ideas we consider now, the more likely we are to find the right ones.[2]

This chapter offers a guide to the rest of this volume. Section I describes the problem of global climate change and the challenges of addressing this complex phenomenon in an economically uneven and politically divided world. The section gives particular attention to the tensions between industrialized and developing countries in the climate change negotiations. Section II probes the key architectural elements that could collectively constitute an international climate protection architecture. These elements include, among others, the legal character of commitments (binding or non-binding), the type of greenhouse gas limitation commitment (emission target or tax), the scope of the action (sectoral, national, or global), and the use of market mechanisms. The chapter concludes with Section III, which offers short summaries of the different approaches to climate protection examined in this volume.

Following this chapter are comprehensive examinations of the different approaches. Chapter 2 examines the Kyoto Protocol itself, which is particularly important because this agreement provides the starting point for discussion of future climate protection options. Given its procedural features, the Kyoto treaty can be adapted to accommodate a variety of approaches, including those examined in subsequent chapters of this volume. In its current form, the Kyoto Protocol establishes fixed "caps" on the emissions of industrialized countries but does not include formal emission-limitation commitments for developing countries. Chapter 2 explores the viability of extending Kyoto's system of emission caps to developing countries.

Chapters 3 and 4 examine two new approaches for structuring participation in greenhouse gas emission reductions by developing countries—Sustainable Development Policies and Measures (SD-PAMs) and a Sector-Based Clean Development Mechanism (Sector-CDM), respectively. Chapter 5 examines "dual-intensity targets," an alternative method of designing emission limitations, while Chapter 6 recounts Argentina's unsuccessful attempt to design and implement a voluntary greenhouse gas target. Chapter 7 looks at the Brazilian Proposal, which calls for emission reductions to be shared among countries on the basis of their relative responsibilities for global warming. Chapter 8 examines a system for distributing emission entitlements to countries on an equal per capita basis. It is important to note that these chapters are not a comprehensive cataloging of approaches: there are others not examined here,[3] and indeed others that have yet to be conceived. However, the approaches that are examined collectively encompass a thought-provoking and wide spectrum of future possibilities.

Each of these chapters assesses both the advantages and disadvantages of a particular approach. The authors pay special attention to the international appeal of the respective approaches because any successful climate protection system must be able to garner international consensus. Thus, the analyses presented in this volume explore the underlying factors—such as economic cost, fairness, and development benefits—that ultimately shape whether approaches are politically acceptable. No single approach can be designated as most desirable, and, in fact, some of them are mutually compatible, as several chapters illustrate.

Chapter 9 offers a quantitative comparison of three different approaches to differentiating greenhouse gas limitation commitments across countries. It illustrates how different approaches can deliver widely varying economic results for a given country. In other words, the disparities among countries ensure that no single strategy examined in this study (or even beyond this study) will be in the best interest of all countries. This analysis reinforces the need to examine the political viability of a variety of approaches, as is done in Chapters 2 through 8. The volume concludes with Chapter 10, which summarizes the advantages and challenges of the different approaches, explores some cross-cutting themes, and outlines some lessons that emerge from this study. Chapter 10 also sets out a way forward for the climate negotiations, incorporating many of the advantages of the approaches examined in this volume.

Box 1.1. The Scope of the Global Climate Change Problem

According to the Intergovernmental Panel on Climate Change (IPCC 2001c), climate change is a vastly different problem than other environmental and public policy issues. Below are six characteristics of the problem that help explain why this is so and militate against easy solutions.

The problem is global. Climate change is related to the concentration of greenhouse gases (GHGs) in the Earth's atmosphere. Emissions from all sources from all countries determine the concentration of these gases. Some countries are very large emitters, and others are very small emitters. Acting alone, individuals and countries that reduce emissions will have a small overall effect.

The problem is long-term. Emissions of carbon dioxide (CO_2), on average, remain in the atmosphere for about 100 years (some other gases persist for thousands of years). Thus, GHG concentrations are related to the net accumulation of gases over long periods of time, not to a single year's emissions. This raises complicated ethical questions because the future generations that will be most affected by climate change are not present to participate in today's decisions.

Associated human activities are pervasive. GHG emissions are linked to a broad array of human activities, including those related to energy use, industrial activities, and land use decisions. In addition, the wide range of policies affecting technological innovation, economic growth, and population size further shape emissions.

I. Confronting Climate Change

Addressing global climate change is a paramount challenge of the 21st Century (Box 1.1). Since the beginning of the industrial revolution, atmospheric concentrations of carbon dioxide (CO_2), the chief heat-trapping greenhouse gas, have risen 35 percent—from about 275 parts per million by volume (ppmv) then to 370 ppmv today. This increase is due to human activities, primarily from the burning of fossil fuels and from deforestation. Carbon that has been sequestered in the Earth's crust (in the form of oil, coal, and other fossil fuels) over millions of years has been extracted, burned, and released into the atmosphere in large quantities within the past 200 years. Atmospheric concentrations of methane, the second leading greenhouse gas, have more than doubled over the past two

Box 1.1. *continued*

Uncertainty is pervasive. Many uncertainties exist regarding the magnitude of future climate change and its consequences, as well as the costs, benefits, and barriers to implementation of possible solutions.

The consequences are potentially irreversible and are distributed unevenly. Sea level rise and other potential consequences of a global temperature increase can take more than one thousand years to play out. Likewise, societies differ in their vulnerability to climate change impacts, with poorer societies less able to adapt to the consequences of climate change.

The global institutions needed to address the issue are only partially formed. The 1992 Climate Convention has nearly universal membership (including the United States). This agreement establishes an objective of stabilizing atmospheric GHG concentrations at a level that would avoid "dangerous" human interference with the climate system. The definition of "dangerous," however, is left open to broad interpretation by Parties. The 1997 Kyoto Protocol has expanded the decision-making process for climate change policy, but currently includes only short-term targets for some industrialized countries.

Source: Adapted from Toth and Mwandosya (2001: 606–609).

centuries. These changes in the composition of the Earth's atmosphere have increased the average global surface temperature by about 0.6° C (1° F) over the past 100 years. Regional climate changes due to temperature increases have already affected many physical and biological systems, and emerging evidence suggests impacts on human settlements from recent increases in floods and droughts (IPCC 2001b).

If the trends in greenhouse gas emissions growth are not altered, global temperatures are expected to rise between 1.4 and 5.8° C (2.5 to 10.4° F) by 2100, according to the latest assessment of the Intergovernmental Panel on Climate Change (IPCC 2001a). The effects of such temperature changes on agricultural production, water supply, forests, and overall human development are unknown but will likely be detrimental to a large portion of the world's population (IPCC 2001b). To prevent atmospheric CO_2 concentrations from exceeding a level of 450 ppmv, global emissions would need to decrease dramatically during this century. Over the same period, however, the global population is expected to increase by 40 to 100 percent (from today's population of six billion) and economic growth is projected to climb 10- to 20-fold (IPCC 2000a). The challenge is formidable and unprecedented; meeting it will require a transition away from a global economy dependent on fossil fuels to one based on renewable and more energy-efficient technologies. Even limiting atmospheric CO_2 concentrations to a higher level, such as 550 ppmv, would entail major emission reductions from projected levels and eventual reductions far below today's emission levels.

Climate change is as much an economic and political challenge as a scientific and technological one. The nature of the problem demands a coordinated approach among the world's countries. Governments resist acting alone to rein in their emissions, given that the rising greenhouse gas output in other countries could undermine their own potentially costly efforts. Furthermore, most emissions come from sectors such as electricity generation, transportation, and agriculture, which are important to national security and economic growth. Powerful vested interests in these sectors will make the transition to a low-carbon future an uphill political climb.

International cooperation is most important—and most challenging—between rich and poor countries. Industrialized countries—primarily the United States, but also others, such as Japan and Australia—are concerned that current lack of emission control commitments for developing countries translates into a lack of environmental effectiveness. This concern is due to rising greenhouse gas emissions in poorer countries as well as the possibility that, given asymmetric emission control commitments, some energy-intensive industries might migrate to countries where emissions growth is unconstrained. Figure 1.1 shows that, although expected growth is large in industrialized countries, CO_2 emissions are expected to grow at much faster rates in China, India, Latin America, and other developing regions over the next few decades. Industrialized countries also argue that, through the 1992 United Nations Framework Convention on Climate Change (UNFCCC) and the subsequent Kyoto Protocol, they have made commitments to curb their greenhouse gas emissions *and* provide financial assistance to developing countries, all without any promise of future action from the developing world. While accepting that richer countries must take the largest steps, they argue that developing countries must take—or at least declare an intention to take—smaller steps.

For their part, many developing countries believe that the industrialized countries lack credibility on the issue of international cooperation to curb greenhouse gas emissions, having done little to address a problem largely of their own making. Figure 1.2 shows that industrialized countries are responsible for most of the buildup of atmospheric carbon dioxide over

Figure 1.1. Carbon Emissions in 2000 and Projected Growth

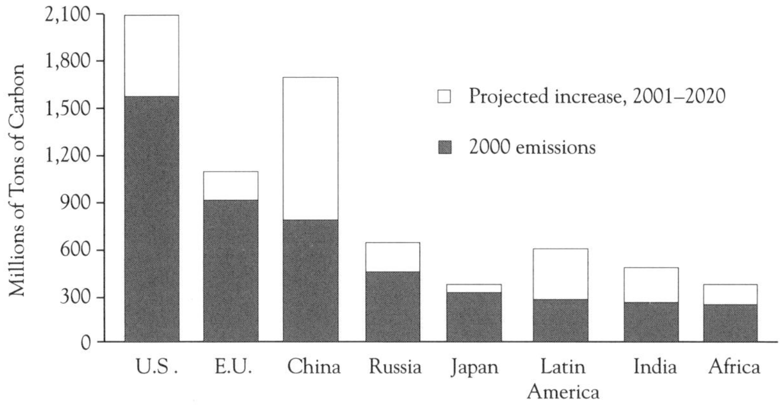

Source: World Resources Institute, compiled from data in EIA (2002a, b).
Notes: Includes carbon emissions associated with fossil fuel combustion; calculations are based on EIA *reference case* scenarios.

the past century created by fossil fuel burning and land use changes (such as deforestation).[4] Specifically, the industrialized countries are responsible for about 63 percent of human-related carbon dioxide that has accumulated in the atmosphere. The 80 percent of the world's population living in developing countries has contributed about 37 percent.

Similarly, although all emissions contribute equally to global warming, large disparities in per capita emission levels reveal a social character of carbon emissions that differs widely from country to country. Figure 1.3 shows that the average American, for example, emits about 10 times more carbon than the average Chinese and 20 times more than the average Indian. Around the world, most people view CO_2 emissions from the United States as resulting largely from luxuries that are unavailable to most people in developing nations, whereas they view the emissions of poor nations as primarily for basic human needs, such as food, warmth, and shelter. Disparities in emissions also reflect an uneven distribution of energy resources throughout the world, with some countries dependent on coal (a fuel that releases relatively large amounts of carbon per unit of energy produced), whereas others rely on less carbon-intensive energy sources, such as natural gas and hydropower.[5]

Figure 1.2. Contributors to Climate Change

Percent of Total Accumulated Atmospheric CO_2 from Industrial Sources and Land Use Changes, 1900–2000

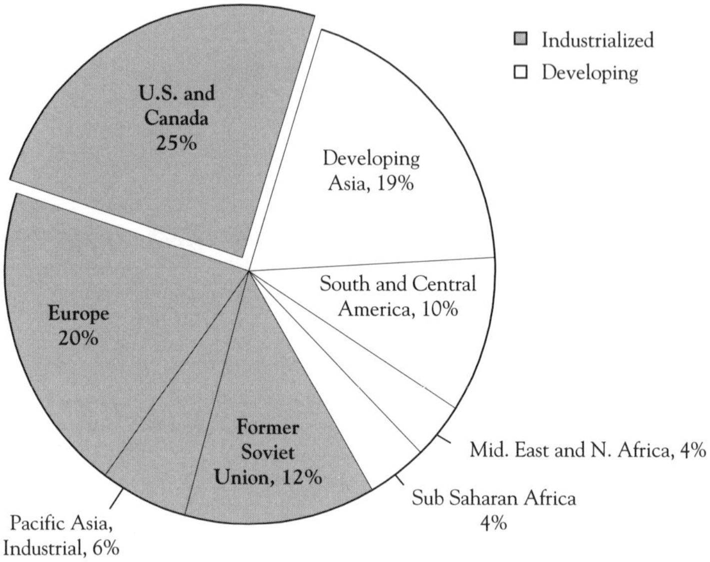

Source: World Resources Institute, compiled from data in Marland et al. (2000), EIA (2002b) and Houghton et al. (2000).

Notes: Data include net CO_2 emissions from fossil fuel combustion (1900–2000), cement manufacturing (1900–1979), and changes in land use (1900–1990), such as harvesting of forest products, clearing for agriculture, and vegetation re-growth.

Many in the developing world feel that some richer countries are fulfilling neither the letter nor spirit of the Climate Convention and subsequent agreements. The Convention calls on countries to "protect the climate system…on the basis of equity and in accordance with their common but differentiated responsibilities and respective capabilities." More specifically, it calls on industrialized countries to "take the lead" in protecting the climate (UNFCCC 1992, Article 3.1). Mindful of these principles and the above-mentioned disparities, all countries formally agreed in 1995 that the first round of legally binding emission controls (to be adopted through a protocol) should *not* include developing countries.[6] This agreement reflects an understanding that the wealthier countries have

Figure 1.3. Carbon Emissions Per Person, 2000

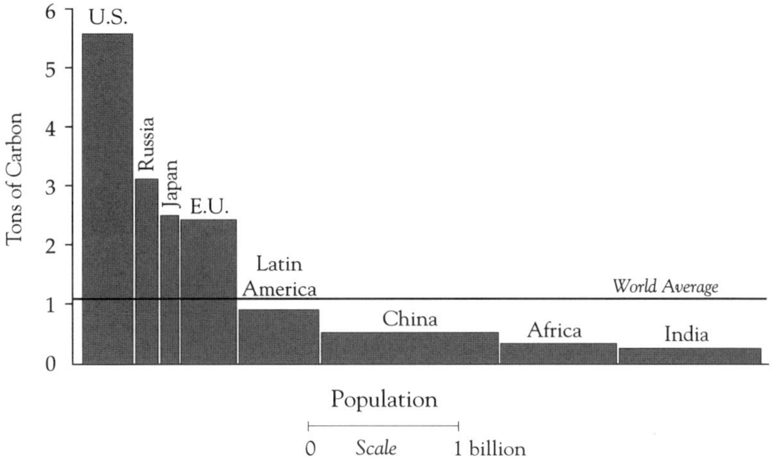

Source: World Resources Institute, adapted from Grubb (1989) and compiled from data in EIA (2002b).
Note: Includes only carbon emissions associated with fossil fuel combustion.

greater financial resources and technological capability to put them on a sustainable course.

Developing countries, on the other hand, face more urgent priorities, such as poverty alleviation and public health. Even when it comes to climate change, the priority for developing countries is generally to reduce their high levels of vulnerability to the physical impacts of climate change, such as sea level rise and extreme weather events. Rather than controlling their emissions, developing countries tend to be concerned about how climate change might have an impact on food production and economic development.

Despite North-South differences in emissions, wealth, and priorities, these disparities are not the largest barrier to cooperation. Probably a more potent obstacle is the enduring and growing lack of trust. Some industrialized countries have legitimate concerns that developing countries may never come into a climate protection regime or may commit only to limit their emissions at some remote future date. For some developing countries, wealthier nations' promises on climate protection seem empty and faithless. Furthermore, some developing countries are concerned that repeated bids on the part of the industrialized countries (principally, but not exclusively, the United States) to include emission limitation commitments for developing countries on the negotiating agenda for the Kyoto Protocol are but thinly veiled attempts to impede poorer countries' economic development prospects. After all, greenhouse gas emissions are intimately linked to essential aspects of economic development, including electric power generation, transportation, and industry. For the developing world, addressing climate change is an issue of basic economic development more than environmental protection.

Recent developments in the Kyoto process could further erode trust and reinforce the North-South stalemate. In March 2001, the United States abandoned the Kyoto Protocol, citing two main reasons: lack of developing-country participation and potentially high economic costs.[7] Yet, the absence of U.S. participation in the Protocol is likely to retard future progress on the very issue that it deemed so important—inclusion of developing countries in an emission limitation regime. In addition to abandoning the Kyoto Protocol, the United States has failed to put a strong climate policy in its place. The policy announced by President Bush in February 2002 will, by the government's own estimates, allow greenhouse gas emissions in the United States to grow by 14 percent from 2002 to 2012 (WRI 2002).

Fortunately, amid the uphill struggle for global cooperation, some grounds for optimism can be reclaimed, even in the United States. In July 2002, the state of California approved a law that will establish the first major greenhouse gas emission standards in the country. Under this law, automakers will be required by the end of the decade to limit greenhouse gas emissions from new cars and light trucks sold in California; such sales account for about 10 percent of total U.S. auto sales. Following this action, 11 additional states sent a letter to President Bush asking for federal measures to limit greenhouse gas emissions.[8]

Elsewhere in the world, most of the industrialized countries, including the members of the European Union and Japan, have ratified the Kyoto Protocol, which is poised to enter into force. With the Kyoto Protocol coming to life, discussions of what comes next gain increasing legitimacy and even urgency. Over the next few years, the pressure for a new round of negotiations will mount. There is already near-consensus that, in the long term, protecting the climate will require controlling emissions from both industrialized and developing countries. The issues are how, when, and under what conditions such emissions will be limited. What comes next?

How might the Climate Convention and Kyoto Protocol be adapted to constitute a more effective multilateral environmental regime? These questions are the subject of this volume.

II. Designing a Climate Protection Architecture

Debates on the future of the climate change treaty have focused overwhelmingly on defining emission targets or, alternatively, on how to allocate future greenhouse gas emission rights across countries. Although they are a central feature of a future climate regime, emission targets are but one part of a coherent climate protection architecture. The approaches to climate protection examined in this volume vary widely with respect to their legal character, geographic scope, use of market-based mechanisms, and other important elements of a climate protection architecture. This section explores each of these elements in greater detail. Table 1.1 summarizes the various elements and options of an international climate protection architecture.

The options discussed in this section could be combined in many ways. Indeed, there are a multitude of permutations for designing a climate protection regime, and especially for distinguishing between developed and industrialized country actions. Just as oil and water do not mix, however, some of these options are incompatible.[9] For example, the right to participate in international emissions trading is conditioned on a country's assuming an emission target with associated monitoring, reporting, and review obligations. Without a balance and coherency to rights and obligations, the environmental integrity of the framework would be sacrificed. As explained in Chapter 2, a proper balance of rights and obligations is a strength of the Kyoto Protocol—access to international emissions trading is restricted to industrialized countries that have also submitted to a battery of other treaty obligations.

Legal Nature of Commitments

Among observers and analysts, confusion often exists over whether a promised action is voluntary or mandatory in international agreements. Generally, all international treaty commitments are made voluntarily, in the sense that sovereign states themselves decide whether to participate in the agreement.[10] Once the treaty comes into force, specific commitments may or may not be considered legally binding.[11] In actuality, the legal nature of a given commitment will probably fall somewhere along a continuum between legally binding and non-binding, depending on the speci-

Table 1.1. Designing a Climate Protection Architecture: Possible Elements and Options

Element of Architecture	Options
Legal Nature of Commitment	1. Legally binding 2. Non-binding
Type of GHG Limitation Commitment	1. International carbon tax (e.g., $10 per ton) 2. Internationally harmonized policies and measures 3. Fixed emission target: cap on emissions (Kyoto-style targets) 4. Dynamic emission target: limit of emissions in relation to GDP growth 5. Dual emission targets: "safe zone" between a high and a low target 6. Emission target with cost cap: target expands if emission reduction costs reach a certain threshold (e.g., $100 per ton) 7. Sustainable development policies and measures (not harmonized)
Coverage and Scope of Actions	1. Gases (e.g., CO_2 only or all six principal greenhouse gases) 2. Geographic (e.g., project, sector, national, regional, global)
Timing and Triggers	1. Determined by existing Annex (e.g., Annex I of Climate Convention) 2. New thresholds for participation: A certain level of income or emissions per capita, for example, determines when a country should take an action
Approach to Differentiating GHG Commitments	1. Pledge-based: Kyoto-style negotiations 2. Principle-based: Agree first on principles and then derive subsequent allocation rules from those principles

ficity of the promised action, the consequences of non-compliance, and the intentions of governments making the agreement. It is worth bearing this continuum in mind throughout this volume.

The climate regime, as explained by Depledge in Chapter 2, currently employs both non-binding and binding commitments. Some provisions, such as the greenhouse gas commitments under the 1992 Climate Convention, are widely considered non-binding pledges.[12] This is due to the general phrasing of the requirements and the lack of an accompanying system of enforcement. The 1997 Kyoto Protocol, however, establishes legally binding requirements for emission limits in industrialized countries. These emission limits are precisely spelled out in the agreement and backed by procedures and mechanisms (adopted in 2001) aimed at remedying cases of non-compliance, such as when a country exceeds its emission limit.

Table 1.1. *continued*

Element of Architecture	Options
Market-Based Mechanisms	1. Project- or sector-based trading (e.g., Clean Development Mechanism) 2. International emissions trading (e.g., Kyoto-style allowance trading)
Financial and Technology Commitments	1. Funding for adaptation, renewable energy investment, sustainable development policies and measures, technology transfer, etc. 2. Compensation for climate impacts
Accountability Commitments	1. Non-compliance consequences 2. Measurement, reporting, review
Environmental Objective	1. Climate Convention objective 2. Agreement to keep a certain stabilization option open in the future 3. A quantitative objective, such as a limit on global emissions, concentrations, or temperature change that is consistent with the Climate Convention objective

Note: Other potential elements of an international climate-protection architecture exist, but are not examined here. Likewise, there are other non-treaty-based strategies, such as technology-driven approaches, which are not examined here.
Abbreviations: GHG (greenhouse gas), GDP (gross domestic product).

Like other environmental agreements, the climate change regime might successfully incorporate non-binding approaches (beyond those already stipulated in the Convention) into its architecture.[13] Several approaches discussed in this volume—SD-PAMs (Chapter 3), Sector-CDM (Chapter 4), and dynamic targets (Chapter 5)—might be quite effective in a non-binding form. Past experience with the Climate Convention, however, suggests that a *purely* non-binding system is unlikely to prevent dangerous climate change.[14] In the future, the climate regime could adopt mandatory requirements for all countries or, instead, for a subset of countries, such as those with greater responsibility for the problem of climate change or those with greater capabilities to reduce emissions.

Type of Greenhouse Gas Limitation Commitments

Greenhouse gas limitation commitments will form a central element of a future climate protection architecture. Here, policymakers have a variety of options. Some commitments would entail *harmonized* policies and measures across countries, such as the removal of fossil fuel subsidies or promotion of renewable energy. Countries could likewise promote climate protection through an international carbon tax. Although theoretically appealing, these kinds of internationally harmonized approaches have had limited traction in climate negotiations over the past decade. Such proposals suffer from monitoring and enforcement problems and, perhaps most important, tend to intrude into the domestic policymaking sphere in a way that has proven politically unacceptable.

Emission targets offer several benefits relative to harmonized actions. First, by their very nature, targets can be differentiated across countries. The Kyoto Protocol targets, for example, range from a 10 percent increase above 1990 levels (Iceland) to an 8 percent reduction below 1990 levels (European Union and others). The concept of differentiation, rather than harmonization, better reflects the Climate Convention's promise to give "full consideration" to the "specific needs and circumstances" of Parties (UNFCCC 1992, Article 3.2). Second, decisions on *how* to achieve emission targets are left to the sovereign discretion of countries, without the intrusion of international rules.[15] Generally, the preference for emission targets (and trading, discussed below) is due to the legal framework underpinning international agreements, which is based on sovereignty and therefore voluntary assent (Wiener 1999). In effect, voluntary assent makes harmonized approaches—such as a global carbon tax or internationally coordinated policies—less politically workable than targets. Through a structured negotiating process, countries commit to a target they find politically acceptable with respect to environmental stringency and economic costs. Third, emission targets are compatible with market mechanisms (see below) such as international emissions trading, which can help reduce overall costs.

One kind of target is a *fixed* (or, absolute) *target*, which establishes a maximum level of emissions a country can emit during a specified period. For example, targets taken by industrialized countries under the Kyoto Protocol entail fixed emission ceilings during the 2008 to 2012 time frame (Chapter 2). Fixed targets have the advantage of ensuring a particular environmental outcome (via a "cap" on emissions)[16] and can promote cost-effectiveness when coupled with emissions trading.

The difficulty with negotiating fixed targets stems from uncertainties over future emission levels and the costs of achieving any future emission target (Baumert et al. 1999, Pizer 1999, Victor 2001). The further into the future targets are set, the greater the uncertainties. These uncertainties carry two opposing risks: (1) a target set too stringently can potentially constrain economic development, an unacceptable consequence for many developing countries and (2) a target set too loosely, in contrast, can result in a weakening of other countries' targets. This second risk is due to the influence of international emissions trading: for instance, country A's excess emission allowances (due to weak targets) might be traded to country B, which would, as a consequence, be able to increase its own emissions. (This phenomenon is often referred to as "hot air" trading.)

At least three ways of designing emission targets could potentially reduce economic uncertainties and environmental risks. The first is a *dynamic target*. Under this kind of target, a country's allowable level of emissions is adjusted according to some other variable, such as gross domestic product (GDP) (CCAP 1998, Baumert et al. 1999, Philibert and Pershing 2001). A dynamic target of this sort was proposed by Argentina in 1999 (Chapter 6). Dynamic targets can reduce economic uncertainty in the target-setting process and promote environmental integrity (i.e., less unintentional "hot air"), particularly with respect to developing countries. Yet, dynamic targets pose certain challenges relative to fixed targets, including added complexity and data requirements. These challenges are explored in greater depth in Chapter 5.

A second way to design emission targets is to use *dual targets*. Here, a country has two emission targets, rather than one. The purpose of the lower (more stringent) target is to provide an incentive to reduce emissions, since reductions below this target would enable the country to sell emission reduction allowances. The higher (less stringent) target would have a punitive function: Exceeding this target puts the country out of compliance. Thus, the lower target would be a selling target and the higher one a compliance target. No penalty would be assessed if emissions fell between the selling and the compliance targets. That area would be the safe zone, in which the country is neither out of compliance nor able to sell allowances through international emissions trading. This dual target concept, as explored in Chapter 5, could be combined with a dynamic target approach.

The third way addresses cost uncertainties by coupling a fixed target with a *cost cap*, sometimes referred to as a "safety valve" or "price cap" (Pizer 1999, Victor 2001).[17] A cost cap places an upward limit on the costs of emission reductions, thereby providing greater up-front certainty about the potential magnitude of implementation costs for a given target. If abatement costs exceed the cap (e.g., \$100 per ton of CO_2), the government may issue additional emission allowances (or purchase them from a central authority), rather than require more costly emission reductions. In such an instance, using the cost cap would allow greenhouse gases to exceed the target level, effectively transforming a fixed target into a dynamic one. Although not explored in any of the chapters, a price cap could work in tandem with several of the proposals examined in this volume.

It should be noted that quantitative emission targets—fixed or dynamic—are not a necessary condition for climate protection, especially for countries whose emissions are relatively small. The 49 countries classified as "least developed" by the United Nations contribute approximately 0.5 percent of yearly global CO_2 emissions.[18] These countries and perhaps others need not necessarily adopt quantitative emission targets or other commitments, even over the next few decades, because their current and future contributions to global greenhouse gas emissions are small.

Moreover, larger developing countries have demonstrated that they can take climate-friendly actions in the absence of firm targets. A wide range of energy efficiency and renewable energy measures are already helping to limit the growth of greenhouse gas emissions in developing countries, even though these measures are being taken for reasons other than climate change (Reid and Goldemberg 1999, Biagini 2000). Thus, *qualitative approaches* that advance country-specific sustainable development policies and measures (SD-PAMs, Chapter 3) could play an important role in developing countries' future climate protection efforts.

Coverage and Scope of Actions

Future commitments could vary with respect to their coverage and scope. Kyoto-style targets, for example, are nearly comprehensive in their emission coverage. They encompass all emission sources and certain sinks (i.e., emission absorption activities) within a country and also address all six main greenhouse gases (carbon dioxide, methane, nitrous oxide, hydrofluorocarbons, perfluorocarbons, and sulfur hexafluoride). In the future, emission limits for some countries could be narrower in coverage, especially in developing countries where some gases are difficult to measure or monitor and may constitute only a small share of countrywide emissions.

A treaty could promote action at the project, sector, or countrywide levels. Project-based emission reductions (which also have sustainable development benefits in developing countries) are already authorized through the Kyoto Protocol's Clean Development Mechanism (CDM). Sector-based commitments might encompass those parts of national economies where greenhouse gas emissions are most prominent, such as heavy industry, land use change and forestry, or electric power production. The approaches examined in Chapters 3 and 4 of this volume—Sector-CDM and SD-PAMs—could be channels for such strategies. Other kinds of international cooperation at the sector level might help address competitiveness concerns, particularly in sectors—such as steel or aluminum—where international competition is often intense.

Internationally, future agreements might entail commitments from a particular region or limited set of countries, such as those that are the largest greenhouse gas emitters. The Kyoto Protocol embodies this approach, in that 38 countries are captured (37, counting the U.S. withdrawal) under the emission control system. It further allows regions, such as the European Union, to achieve their targets jointly through their own internal agreement on commitments. This same strategy could be employed in the future, including in developing countries (see Chapter 2).[19] Finally, a *global* commitment could cover all countries under the same emission control system. Proposals such as those calling for per capita-based emission entitlements fall under this category (Chapter 8).

In general, a regime that is broader in scope and coverage will afford greater opportunity for participants to undertake emission reductions where they are least costly (see market mechanisms below). At the same time, however, the broader the scope, the higher a regime's monitoring and evaluation costs will be.

Timing and Triggers

Future actions could also vary with respect to *timing*: Some countries could take action sooner than others. This kind of differentiation is clearly visible in the Kyoto Protocol, as industrialized countries committed to emission controls from 2008 to 2012, with the prospect of additional countries making commitments in a later period. Since the 1992 adoption of the Climate Convention, action has been differentiated primarily on the basis of countries' designation as "Annex I" or "non-Annex I" Parties, categories that correspond roughly to traditional North-South groupings. Some treaty provisions apply to other categories, including "Annex II" Parties (wealthy countries with a special obligation to provide developing countries with financial and technological assistance on climate change), "economies in transition," "developing" countries, and "least developed" countries (see Chapter 2).

For the future, a challenging issue is determining what should trigger an emission limitation commitment for a particular country. A least developed country might be exempt from greenhouse gas commitments of any kind for several decades, yet a rapidly growing one might not. This issue of when countries should graduate to steeper commitments arises in several of the approaches examined in this volume. Whereas traditional distinctions (i.e., Annex I/non-Annex I) will be useful and necessary in the future, new categorizations may be needed to differentiate the timing of actions across countries. For example, the 1987 Montreal Protocol includes different schedules for phasing out ozone-depleting substances based on a country's per capita consumption of certain controlled substances (0.3 kilograms per person) (Montreal Protocol 2000).

Approach to Differentiating Commitments

As noted above, greenhouse gas emission targets can be differentiated across countries, with some countries required to reduce emissions more than others. It is useful to consider two different procedural approaches to negotiating emission targets, be they fixed, dynamic, or qualitative. The different approaches examined in this volume, as well as others outside the scope of this study, can be categorized as either pledge-based or principle-based. This distinction is important because it determines a starting point for negotiations and, more fundamentally, reflects differing and perhaps conflicting ways of viewing the challenge of climate protection.

Generally, the international negotiating process is best characterized as pledge-based; countries formulate their national positions and negotiate in their interests, voluntarily making commitments (alone or with other countries) at their sovereign discretion. Because the international legal order lacks the ability to require a country to participate, the tradition has been for countries to "pledge" particular actions in a bottom-up style. These commitments typically represent (and always purport to make) some divergence from the status quo. (See Box 1.2.) In the Kyoto Protocol negotiations, for example, industrialized countries pledged various emission limitation or reduction targets relative to their 1990 emission levels. This pledge-based approach reflects the voluntary assent rule and the *realpolitik* of international negotiations.

Bottom-up negotiation processes like the Kyoto Protocol have been criticized as ad hoc, with negotiated results shaped mainly by political power and economic might rather than by objective criteria. Thus, many have called for negotiation on overarching principles or rules that, once agreed, would guide the subsequent emission reduction efforts among nations in an orderly fashion. The Brazilian Proposal (Chapter 7) and per capita allocations (Chapter 8) are two examples of principle-based proposals. The Brazilian Proposal would apportion emission reduction requirements based

Box 1.2. Business As Usual: The Challenge of Setting Targets

In determining emission targets, governments will primarily be concerned with their business-as-usual (BAU, or "baseline") scenario, which represents the most plausible projection of future emissions.[1] BAU embodies the notion of what would happen, hypothetically, if climate-friendly actions were not taken. BAU emission estimates can help governments gauge the stringency and economic acceptability of a particular emission target. The difficulty of estimating BAU patterns stems from trying to forecast the future, a challenge that may be greater in developing and transition countries, which have fewer consistent historical patterns and for which development is more affected by external conditions. Even in mature economies of the industrialized countries, however, accurately predicting future economic and emission trends is difficult.

Dynamic targets (Chapter 5) represent an attempt to lessen some of the problems associated with postulating a BAU reference case by introducing targets that are subject to adjustment according to shifting economic conditions. This was the main appeal of a dynamic target for Argentina (Argentine Republic 1999), which used nine different future scenarios to determine the magnitude and mechanics of the target it ultimately announced. Chapter 5 introduces the concept of "dual targets"—an innovative way to further reduce (but not eliminate) economic uncertainty in establishing emission limitations.

continued on next page

on each country's relative responsibility for the global temperature increase. The per capita approach would distribute allowances according to the size of a country's population. There are other allocation-based approaches—not examined here—that operate similarly.[20]

Principle-based approaches often are advanced under the mantle of *equity*, a stated principle of the Climate Convention. (See Box 1.3.) The most recent IPCC assessment catalogs 13 equity principles and their associated allocation rules (Toth and Mwandyosa 2001), illustrating a diversity of views on what constitutes an equitable allocation of emission allowances across countries. Similarly, many believe that, given North-South disparities in negotiating capacity and power, a principle-based approach to negotiating commitments is fairer *procedurally*. Some developing countries question whether they can ever get a "fair deal" if emission commitments are determined on the basis of raw bargaining power.

Box 1.2. *continued*

The concept of BAU is also relevant to other approaches examined in this volume. The concept of Sustainable Development Policies and Measures (SD-PAMs, Chapter 3) is predicated on taking actions that diverge from a "conventional development path." Emission credits under the Clean Development Mechanism (CDM) or Sector-CDM (Chapter 4) must be "additional" to those that would have occurred otherwise. Both of these approaches involve the same slippery concept of business as usual.

Two of the approaches examined in this volume attempt to sidestep the difficult issue of business as usual, as they are not concerned with the status quo. An allocation based on equal per capita entitlements, or an approach based on relative responsibility for global temperature increase (i.e., the Brazilian Proposal), operates according to predetermined principles that do not involve BAU forecasting. Nevertheless, these approaches do not entirely avoid the morass of BAU. Countries will still be concerned about assessing the reduction efforts required by a given target, and such assessments typically involve forecasts of future emission levels. The political acceptability of a particular target is strongly influenced by the magnitude of emission reductions required, as well as the associated costs or benefits.

---

[1] Even though Kyoto targets are expressed relative to a base year (e.g., 8 percent *below* 1990), the difficulty of meeting them typically involves assessing current emission levels relative to future emission projections (e.g., a 2010 target might amount to a 20 percent reduction compared to projected levels).

To be sure, however, the issue of equity should not be associated solely with the differentiation of emission commitments achieved through principle- or pledge-based negotiations. Equity is relevant to all elements of the architecture. For example, the Montreal Protocol on Substances that Deplete the Ozone Layer is widely perceived as fair not only because country commitments were differentiated but also because industrialized countries ultimately compensated developing countries for phasing out ozone-depleting substances (Banuri et al. 1996). Developing countries agreed to phase-out schedules only after industrialized countries provided the necessary financing through a multilateral fund. Similarly, the level of acceptable climate change (see "Global Environmental Objective" below) will have a major bearing on equity because the impacts of climate change will be unevenly distributed. As suggested by the definition in Box 1.3,

Box 1.3. Equity and Climate Protection

The Climate Convention stipulates that countries should protect the climate system "on the basis of *equity*, and in accordance with their common but differentiated responsibilities and respective capabilities" (emphasis added). In general terms, the Intergovernmental Panel on Climate Change (following on Flexner 1997) defines equity as "the quality of being fair or impartial" or "something that is fair or just." In the narrower context of international environmental issues, Harris defines equity as "a fair and just distribution among countries of benefits, burdens, and decision-making authority associated with international environmental relations." Several equity considerations are embedded in these definitions, including the following:

Procedural equity concerns the fairness of the negotiating process. Albin states that during negotiations all Parties should have the opportunity for fair hearing, fair input, fair play, and fair procedures. More specifically, all Parties should be given full and equal opportunity in the debates and all Parties should negotiate in good faith, reciprocate, and adhere to agreed rules. In addition, decision-making should be transparent and votes representative.

Consequentialist equity refers to the fairness of outcomes or distributive justice, that is, the distribution of greenhouse gas emission limitations and their associated costs and benefits, as well as the burdens associated with adapting to or bearing the physical impacts of climate change. Consequentialist equity also has a temporal dimension: *Intergenerational* equity suggests that actions to protect the climate system are called for in the near term so that future generations do not suffer from unacceptable climatic changes.

Sources: Banuri et al. 1996 (IPCC, above), Albin 2002, Harris 2000.

the totality of the climate protection architecture, not just one element, ultimately will influence whether governments perceive an agreement as fair. Likewise, achieving an internationally acceptable differentiation of greenhouse gas commitments is not just a matter of agreeing on equity principles. Countries may hold fundamentally different worldviews on climate change encompassing very different notions about the urgency of climate protection and the nature of appropriate management strategies (Rayner 1994, cited in Banuri et al. 1996).

Market Mechanisms

Market mechanisms, such as international emissions trading, are increasingly embraced by the international community in efforts to address climate change.[21] The primary attraction of market mechanisms is *cost-effectiveness*, a principle enshrined in the Climate Convention (Article 3.3). Emissions trading supports this principle by providing incentives for emission reductions to be undertaken where they are the least costly. The merits of emissions trading with respect to cost-effectiveness are extensively documented in the literature.[22]

The basic mechanics of international emissions trading are relatively simple. First, governments must commit to emission limitation targets (discussed above). Second, such targets are divided into discrete, tradable units. These tradable units are often referred to as *allowances*, because they "allow" the holder to emit a specified amount of greenhouse gases, say, one ton of carbon dioxide or the equivalent amount of another greenhouse gas. Governments may choose whether to distribute these allowances to domestic emitting sources. Third, allowances could then change hands in several ways—in trades between governments, between a governmental and a private entity, and between private entities. The party purchasing allowances is entitled to emit more; the party selling those allowances is required to emit less.

As explained in Chapter 2, the Kyoto Protocol incorporates an international emissions trading system, as well as two project-based market mechanisms—joint implementation and the CDM. In addition to lessening the cost of greenhouse gas emission cuts, the CDM aims to promote sustainable development in the developing world. Several other approaches discussed in this volume could also use market mechanisms. SD-PAMs (Chapter 3) could entail access to the CDM. Sector-CDM (Chapter 4) suggests expanding the scope of the CDM to encompass entire sectors or geographic regions. Dynamic targets (Chapters 5 and 6), variants of the Brazilian Proposal (Chapter 7), and a system of per capita–based entitlements (Chapter 8) could each use an international emissions trading system. Chapter 9 illustrates the cost-effectiveness benefits that could be realized under a well-functioning trading system.

Financial and Technology Commitments

Financial provisions—such as those for capacity building, adaptation assistance, and technology transfer—are essential to crafting North-South compromises. For example, the final package adopted as the 2001

Marrakesh Accords comprises technical provisions for making the Kyoto Protocol operational *and* a financial component, however limited, aimed at helping developing countries address climate change and adapt to its physical impacts. This financial package includes an adaptation fund and a least developed country fund for which industrialized countries have pledged a relatively small amount of money.[23] Currently, the climate change regime has designated the Global Environmental Facility (GEF) as its financial mechanism. In addition to managing several funds for the Climate Convention, the GEF finances activities supporting the Convention's implementation in developing countries, including capacity building, preparation of national communications and greenhouse gas inventories, and vulnerability and adaptation assessments.

To the extent that developing countries are asked to take on new greenhouse gas commitments, the design and funding of financial mechanisms will be critical. According to the Climate Convention, the degree to which developing countries will effectively implement their commitments depends on the degree to which they receive assistance from the industrialized countries (UNFCCC 1992, Article 4.7). The IPCC further states, "Most analysts...suggest that both equity and efficiency considerations create a case for large international financial transfers as part of any regime for substantial reductions in greenhouse gas emissions" (Banuri et al. 1996).

Accountability Provisions

Any effective climate protection architecture will require provisions for determining whether countries are adhering to their promises. These provisions include national monitoring and reporting as well as the review of information (such as emissions data) submitted by Parties in order to ensure accuracy and completeness. These requirements are essential conditions for implementing some options for greenhouse gas limitation commitments as discussed above, since a government cannot manage what it cannot or does not measure. Equally important are the procedures and consequences to which countries are subject if they fail to comply with their obligations (or are suspected of non-compliance). Such credibility mechanisms are already enshrined in the Kyoto Protocol and the subsequent 2001 Marrakesh Accords and will undoubtedly form important building blocks for future climate agreements.

Global Environmental Objective

To what end is the above-discussed climate protection architecture directed? What constitutes "climate protection"? The Climate Convention (and by association the Kyoto Protocol) establishes, as an ultimate objective, the "stabilization of greenhouse gas concentrations in the atmosphere at a level that would prevent dangerous anthropogenic interference with the climate system." The Convention also stipulates that, "Such a level should be achieved within a time-frame sufficient to allow ecosystems to adapt naturally to climate change, to ensure that food production is not threatened and to enable economic development to proceed in a sustainable manner."

There is currently no agreement on what constitutes a dangerous level of greenhouse gas concentrations. The IPCC's most recent assessment report states that, "Given the large uncertainties that characterize each component of the climate protection problem, it is impossible to establish a globally acceptable level of GHG concentrations today" (Toth and Mwandyosa 2001). Yet, an important step to building an environmentally effective climate regime would be to achieve greater clarity on a long-term goal (Berk et al. 2001, Corfee-Morlot 2002). A long-term perspective casts the climate protection challenge into sobering relief: Even limiting atmospheric greenhouse gas concentrations to a *doubling* of pre-industrial levels would likely require a wholesale transition in the world's energy economy. A more formal long-term objective might help shape more effective near-term actions in a way that is consistent with a variety of future atmospheric stabilization options. One promising approach explored by the COOL Global Dialogue Project is to ensure that global commitments keep future climate protection options open to a stabilization of CO_2 concentrations at 450 ppmv (550 ppmv including all gases) (Berk et al. 2001).

III. Seven Approaches to Climate Protection

Rigid divisions between industrialized and developing countries have been a main feature of the international climate debate (and indeed other debates) over the past decade. To protect the atmosphere from dangerous climate change, the coming decade must witness the bridging of this divide between richer and poorer nations. Accomplishing this will require, first and foremost, industrialized country leadership, as called for under the Climate Convention. New ways of designing international cooperation, such as those examined here, will also need to be considered, if not adopted.

When examining the different approaches, here and in subsequent chapters, it is important to avoid false comparisons. The approaches presented differ in scope, application, and purpose; they represent different elements of a climate protection architecture. Some of them are mutually compatible, and none of them represents the best or most appropriate approach for all countries. Chapter 10 explains the elements of a climate protection architecture that are prominent in each approach.

The Kyoto Protocol

In Chapter 2, Joanna Depledge describes the salient features of the climate protection architecture that currently exist in the 1997 Kyoto Protocol. As the focal point of international negotiations from 1995 to the present, the Kyoto treaty is clearly the architecture with which the international community is most familiar and comfortable. No doubt, some elements of the Protocol will persist for decades to come, and other approaches, such as those described in this volume, may even be absorbed into the Kyoto Protocol framework in the future. Depledge assesses the viability of continuing Kyoto—preserving the existing architecture (e.g., fixed targets) but widening its scope to include developing countries. The author also examines the processes through which the Kyoto Protocol can expand and embrace some of the other approaches presented in this volume. In the process of articulating the Kyoto Protocol's architecture and its procedures, Depledge illuminates the historical context of the current tensions between industrialized and developing countries.

Sustainable Development Policies and Measures

In Chapter 3, Harald Winkler and his colleagues examine Sustainable Development Policies and Measures, an innovative approach for developing countries to contribute to climate protection. For most developing countries, climate change is not an immediate priority, and sustainable development could be a more robust objective around which to organize action. The SD-PAMs approach harnesses this reality by beginning with the development objectives and needs of developing countries. If countries act early to move to greater sustainability in their development path, they will start bending the curve of their greenhouse gas emissions downward. The approach's logic is derived directly from the Climate Convention, which states that countries should promote "sustainable development" through "policies and measures to protect the climate system" (Article 3.4). As the authors point out, the emphasis in the climate negotiations has been to focus on emission reductions rather than sustainable development. The approach is illustrated through a case study of South Africa.

Sector-Based Clean Development Mechanism

Chapter 4, by Joséluis Samaniego and Christiana Figueres, examines a *sector*-based Clean Development Mechanism and considers its potential application in Mexico City. This approach builds on the already operational *project*-based CDM. Currently, CDM rules and institutions are designed primarily to encompass projects that are relatively narrow in scope, such as electric power or energy efficiency projects. The Sector-CDM represents an expansion of the scope of the CDM to cover entire national sectors (such as cement or power production) or geographic areas (such as a municipality). This approach could support emission reductions and sustainable development benefits—the two expected by-products of the CDM—across a wider array of activities. This approach could also bring financial resources to fund the kind of SD-PAMs discussed in Chapter 3. It is important to note that this approach differs from others in this volume in that a Sector-CDM is not necessarily a post-Kyoto one. Because the definition of a CDM project is currently indeterminate, new rules conceivably could make this approach operational in the relatively near future.

Dual-Intensity Targets

In Chapter 5, Yong-Gun Kim and Kevin Baumert explore two distinct ideas—dynamic targets and dual targets—and how, individually and in combination, they can be used to reduce the uncertainties inherent in committing to greenhouse gas emission limitations. Establishing greenhouse gas targets is a contentious process. Future emission levels are highly uncertain (especially in developing countries), and countries are wary of any commitment that could turn out to be excessively stringent. Yet, weak targets might not deliver any environmental benefits. Dynamic targets, as alluded to in Section II of this chapter, could help address these challenges. Dynamic targets differ in two important respects from the fixed targets adopted in the Kyoto Protocol. First, dynamic targets adjust according to a variable (such as GDP) that typically has a strong influence on emission levels. The adjustments could be made by using *intensity* or *indexed* targets—the two main kinds of dynamic targets. Second, more than one target can be used. In this regard, Kim and Baumert explore the viability of using high- and low-intensity targets (i.e., dual targets), between which a country would occupy a "safe zone" where it would neither be out of compliance nor able to sell emission allowances.

Argentine Voluntary Commitment

In Chapter 6, Daniel Bouille and Osvaldo Girardin offer a specific and unique example of a country that sought, and failed, to implement a voluntary commitment based on a dynamic target. In 1998, Argentina declared its intentions to be bound by an emission limitation target and further elaborated on the specifics of that target the following year. The authors illuminate the political context under which Argentina made its commitment, emphasizing Argentina's desire to align its foreign policy objectives with those of the United States. The chapter likewise demonstrates that target setting is a technically complex exercise, in part because of the uncertainty of future emissions and the treatment of gases other than CO_2, such as methane from agriculture. The chapter articulates several lessons learned from the Argentine experience—both in deciding to commit to a voluntary target and in specifying the target's nature and level—that could be useful to other countries contemplating similar actions.

Brazilian Proposal

In Chapter 7, Emilio La Rovere, Laura Valente, and Kevin Baumert explore the Brazilian Proposal. The most salient feature of this Proposal is its call for sharing emission reduction burdens on the basis of each country's relative responsibility for the global temperature increase. This idea is derived from the "polluter pays" principle and builds squarely on the Climate Convention language calling for all countries to protect the climate system according to their "common but differentiated *responsibilities* and respective capabilities" (Article 3.1, emphasis added). The Brazilian Proposal is the only approach examined in this volume that has been officially proposed to the UNFCCC Parties. The proposal was originally submitted in July 1997, before the adoption of the Kyoto Protocol, and revised in 1999. Although the Proposal played an important role in the Kyoto negotiations, it did not garner a consensus. It may yet play an important part in future debates on shaping a global climate protection system. Chapter 7 explores various ways that the Proposal could be expanded to include developing countries and adapted to increase its workability.

Per Capita Entitlements

In Chapter 8, Malik Amin Aslam explores per capita-based emission entitlements, the approach that has gained perhaps the most attention of any examined in this volume. The per capita entitlements idea is based on the notion of "equitable" resource sharing and elaborates on the proposition that each person has the right to emit an equal amount of greenhouse gases. Although there are many variants of this approach, most begin by suggesting that overall global emissions must contract to a level that prevents dangerous climate change. Moreover, emissions *per person* must converge from today's levels to one that is equal across all countries. The analysis addresses key issues likely to shape the acceptability and effectiveness of an equal per capita entitlements approach, including issues of equity and the application of various fairness criteria, the importance of international emissions trading, and the ability to account for diverse national circumstances. The author proposes an alternative that might increase the political appeal of a per capita-based solution.

Notes

1. The Kyoto Protocol's first commitment period runs from 2008 to 2012. Although this may sound far away, negotiations for the second commitment period must be concluded by the end of 2007, according to the Protocol's own provisions. Negotiations could begin between 2003 and 2005.
2. See Menand (2001, 431) for a metaphor of probabilistic thinking.
3. Indeed, some strategies to protect the global climate system do not even involve an international treaty. This volume does not address such strategies.
4. If non-CO_2 gases could be included in this figure, the responsibility would shift somewhat toward developing countries because many developing countries have a higher share of non-CO_2 gases than industrialized countries.
5. Some hydropower installations can result in significant emissions of greenhouse gases, particularly dams in tropical countries. See WCD (2000).
6. This agreement, known as the Berlin Mandate, establishes that the protocol or other legal instrument to be negotiated should not include new commitments for developing countries.
7. For treatment of this subject, see Baumert and Kete (2001).
8. For more information, see the Natural Resources Defense Council webpage at http://www.nrdc.org/media/pressreleases/020722.asp.

9. Some of the elements of a climate protection architecture address the rights or privileges of governments (or their private entities) under the treaty. These include the right to participate in international emissions trading and the CDM. Other elements of differentiation address the obligations of countries—what they should or must do. Access to particular treaty rights typically comes part and parcel with the assumption of parallel obligations.
10. "Voluntary" is an ambiguous term. For this reason, we use the term "non-binding" instead. See Werksman (1999) for a discussion on this point.
11. The case of Argentina exemplifies this distinction (Chapter 6). Although its commitment was dubbed "voluntary," Argentina's communication to the UNFCCC stated that "the present commitment shall constitute a binding international commitment once the Conference of the Parties to the UNFCCC implements a new option that may enable non-Annex I Parties" to assume a target and participate in the mechanisms (Argentine Republic 1999). If the COP had taken such action, Argentina's target would be considered voluntary and binding.
12. Experts disagree on this point. Literally speaking, treaty commitments are binding on Parties. However, the loose phrasing of commitments in the UNFCCC, according to many, renders them aspirational rather than mandatory commitments. Article 4.2b, for example, requires industrialized countries to "communicate... detailed information on its policies and measures... with the aim of returning individually or jointly to their 1990 levels these anthropogenic emissions of carbon dioxide and other greenhouse gases." This differs from Kyoto's clear emission reduction requirements and non-compliance procedures.
13. For example, see Levy et al. (1992). When viewed as legally binding regulatory rules, many agreements will appear as ineffective (e.g., Convention on Long-Range Transboundary Air Pollution, or LRTAP). Yet, multilateral environmental agreements can establish important norms and principles, increase governmental concern for environmental problems, and catalyze processes that enhance the capacity of governments to address environmental problems.
14. Many industrialized countries have not achieved the emission limitation pledges made in the Climate Convention.
15. Another reason for the preference for trading over taxes (i.e., quantity over price instruments) identified by Pizer (1999) is the hope that emission allowances might be distributed free, whereas a carbon tax would require a transfer of revenue to governments.
16. Of course, this assumes that countries comply with the target.
17. Like a carbon tax, the level of a cost cap would need to be harmonized across countries.
18. Authors' calculations, based on EIA (2001a).
19. The European Union, however, is still a special case because both the member states and the European Union are Parties to the Protocol. The European Union has strong institutions that can exercise jurisdiction over its members in important policy areas. Thus, the joint fulfillment of commitments among a regional grouping will be harder in other cases.

20. For other principle-based proposals, see, specifically, Blanchard et al. (2001), Groenenberg et al. (2001), Gupta and Bhandari (1999), and Müller (2001b), and, generally, Banuri et al. (1996) and Toth and Mwandosya (2001).
21. Taxes, of course, are also a kind of market-based mechanism. As discussed above, however, greenhouse gas taxes are more likely to be implemented at the domestic than international level.
22. See Baumol and Oates (1988) and Tietenberg (1985). For experiences with U.S. domestic programs, see Carlson and Burtraw (2000), Stavins (2001), and US EPA (1985). At the international level, however, emissions trading is relatively untested. Achieving the positive results that have been demonstrated in domestic contexts will require competitive markets and other conditions that may prove elusive, especially within the confines of international treaty law where participation and compliance cannot be assured (Baumert et al. 2002). The Kyoto Protocol constitutes the first major experiment in international emissions trading.
23. UNFCCC 2002, Decisions 5/CP.7 and 7/CP.7.

2. CONTINUING KYOTO: *Extending Absolute Emission Caps to Developing Countries*

Joanna Depledge

Introduction

The climate change regime, consisting of the 1997 Kyoto Protocol and its "parent" treaty the 1992 United Nations Framework Convention on Climate Change (UNFCCC), provides the foundation for international efforts to address climate change. (See Box 2.1 for a brief history of the climate change regime.) Under this regime, while all countries have general obligations to address climate change, only industrialized countries are subject to specific emission controls, with legally binding caps on their greenhouse gas (GHG) emissions established in the Kyoto Protocol. However, the ultimate objective of the Convention and the Kyoto Protocol—to prevent dangerous human interference with the climate—cannot be met without also entailing specific controls on the emissions of developing countries. Although the climate change regime is silent on the nature of such controls and how they should be introduced, the default path would be to continue the Kyoto Protocol's approach to emission controls, that is, for developing countries to also assume legally binding caps on their emissions. The challenges posed by "continuing Kyoto" in this way make up the focus of this chapter.

The chapter also draws attention, however, to the fact that the procedural framework of the climate change regime is sufficiently flexible to be able to accommodate *any* of the alternative approaches to emission controls explored by the Climate of Trust project. Moreover, some of the substantive mechanisms of the Kyoto Protocol—such as its reporting system and even emissions trading—could also be applied to other forms of emission controls. Kyoto could therefore also continue under a different emission control system, not just one of absolute emission caps. Whatever system is chosen, continuing the framework of Kyoto would have the great

Box 2.1. Major Milestones in the Climate Change Regime

1988	UNEP and WMO establish the Intergovernmental Panel on Climate Change (IPCC).
1990	The UN launches negotiations on a framework convention on climate change.
1992	The UN Framework Convention on Climate Change is adopted in New York and opened for signature at the Earth Summit in Rio de Janeiro, Brazil. The Convention receives 154 signatures and enters into force in 1994.
1995	- The first Conference of the Parties (COP 1) in Berlin launches a new round of negotiations to strengthen targets for Annex I Parties. - IPCC second Assessment Report concludes that the balance of evidence suggests a discernable human influence on the global climate.
1997	COP 3 meeting in Kyoto, Japan adopts the Kyoto Protocol.
1998	COP 4 meeting in Buenos Aires, Argentina adopts the "Buenos Aires Plan of Action," setting out a program of work on the Kyoto Protocol's operational rules and the implementation of the Convention. The deadline for achieving these rules is set for 2000.
2000	COP 6 meets in The Hague, but negotiations break down.
2001	- *January*. IPCC Third Assessment Report is released. - *March*. U.S. President George W. Bush announces that the United States will not become a Party to the Kyoto Protocol. - *July*. At the resumed session of COP 6, Parties adopt the "Bonn Agreements," a political deal on the Kyoto Protocol's rules and the implementation of the Convention. - *November*. COP 7 in Marrakesh adopts the "Marrakesh Accords," a set of detailed rules for the Kyoto Protocol and the implementation of the Convention.
2002	- The World Summit on Sustainable Development meets in Johannesburg, South Africa, to review progress since the 1992 Earth Summit. - COP 8 is held in New Delhi, India.

Abbreviations: COP (Conference of the Parties; the annual meeting of Parties under the Climate Convention); UNEP (United Nations Environment Programme); WMO (World Meteorological Organization, IPCC (Intergovernmental Panel on Climate Change)

advantage of building on an architecture that is already in place and understood. The difficult and complex negotiations that were needed to agree the Kyoto Protocol and, later, the rules for its implementation set out in the 2001 Marrakesh Accords, should serve as a warning against any move to negotiate a whole new framework.

The chapter first describes the climate change regime and then analyzes the procedures in place for its further development, both through individual accession by developing countries to absolute emission caps, as well as the launch of a comprehensive new negotiating round. The chapter then considers the advantages and shortcomings of continuing Kyoto through the extension of absolute emission caps to developing countries. Finally, the chapter draws conclusions that serve as a launch pad for examining the alternative options discussed in the remainder of this volume. I.

I. The Climate Change Regime

Principles and Categories of Parties

The climate change regime enshrines a deal that was struck during the negotiation of the Convention on how to distribute the burden of addressing climate change. This deal is founded on the principles of "common but differentiated responsibilities" and industrialized country leadership (UNFCCC 1992, Article 3.1), which have precedents in many other international environmental agreements. The Convention reflects an understanding that, while all countries have a common responsibility to address global climate change, the industrialized countries have a special duty to take the lead, due to their greater historical contribution to climate change, their generally higher per capita emissions, and their more abundant financial and technological resources to respond to the problem. This does not exempt developing countries from action; the assumption is that, once industrialized countries have taken the lead, developing countries will follow. By ratifying the Convention, its 186 Parties have accepted the principles of "common but differentiated responsibilities" and industrialized country leadership, and their application to subsequent agreements in the climate change regime, such as the Kyoto Protocol.

The principles of "common but differentiated responsibilities" and industrialized country leadership broadly underpin the regime's classification of the world's states. This classification is carried over to the Kyoto Protocol from the Convention, which lists 41 countries in its Annex I, calling these "Annex I Parties." The remainder of the world's states, mostly the developing countries, fall into the category known as "non-Annex I Parties." Although, as discussed below, there are procedures in the Convention to enable non-Annex I Parties to "graduate" to Annex I status, in practice this has proved problematic, and only a handful of countries have graduated in this way. The division between Annex I and non-Annex I Parties has thus become rigid, and increasingly fails to reflect the diversity of national circumstances.

Table 2.1. Annex I Parties, Subcategories, and Targets

Annex I Party	Sub-category	Target
Australia	Annex II	+8
Austria	Annex II	–8 (–13)
Belarus	*EIT*	*
Belgium	Annex II	–8 (–7.5)
Bulgaria	*EIT*	–8
Canada	Annex II	–6
Croatia	*EIT*	–5
Czech Republic	*EIT*	–8
Denmark	Annex II	–8 (–21)
Estonia	*EIT*	–8
European Community	Annex II	–8
Finland	Annex II	–8 (0)
France	Annex II	–8 (0)
Germany	Annex II	–8 (–21)
Greece	Annex II	–8 (+25)
Hungary	*EIT*	–6
Iceland	Annex II	+10
Ireland	Annex II	–8 (+13)
Italy	Annex II	–8 (–6.5)
Japan	Annex II	–6
Latvia	*EIT*	–8
Liechtenstein	–	–8
Lithuania	*EIT*	–8
Luxembourg	Annex II	–8 (–28)

Annex I Parties

To date, the Annex I Parties consist of the 24 developed countries that were members of the Organization for Economic Cooperation and Development (OECD) in 1992 when the Convention was adopted, along with the European Community, Liechtenstein, Monaco, and 14 "countries with economies in transition" (EITs), that is, the more industrialized countries of the former Soviet Union and Central and Eastern Europe (see Table 2.1).

The list of Parties in Annex I was drawn up based on membership of political groups—the OECD and, more loosely, the former Soviet bloc—rather than any objective indicator. Even the use of OECD membership proved controversial, with Turkey, an OECD member, arguing that its lower historical emissions and less advanced economy did not warrant its inclusion in Annex I. Moreover, as there is no reference to OECD membership in the Convention, this is not currently an accepted criterion for graduation to Annex I. While Mexico and South Korea have become members of the OECD since 1992, they have not joined Annex I.[1]

Table 2.1. *continued*

Annex I Party	Sub-category	Target
Monaco	–	–8
Netherlands	Annex II	–8 (–6)
New Zealand	Annex II	0
Norway	Annex II	+1
Poland	*EIT*	–6
Portugal	Annex II	–8 (+27)
Romania	*EIT*	–8
Russian Federation	*EIT*	0
Slovakia	*EIT*	–8
Slovenia	*EIT*	–8
Spain	Annex II	–8 (+15)
Sweden	Annex II	–8 (+4)
Switzerland	Annex II	–8
Turkey	*"special circumstances"*	*
United Kingdom	Annex II	–8 (–12.5)
Ukraine	*EIT*	0
United States	Annex II	–7

Notes: Targets represent percentage changes in greenhouse gas emissions during the first commitment period (2008–12) relative to 1990 emissions. Targets in parentheses are individual country targets for the 15 members of the European Union (EU), which redistributed its collective 8 percent reduction target at the E.U. Environment Ministers Council in June 1998.
* Although they are in the Convention's Annex I, Belarus and Turkey do not have targets listed in the Protocol's Annex B as they were not Parties to the Climate Convention when the Kyoto Protocol was adopted.
Abbreviation: Economy in Transition (EIT).

Annex I includes two main sub-categories: the EITs on the one hand, and the OECD members (again, only as of 1992) on the other. The EITs are identified by a footnote in Annex I and are allowed a "certain degree of flexibility" in meeting their commitments.[2] The OECD members, for their part, are also listed in the Convention's Annex II, which means they must provide financial assistance to developing countries and technology transfer to both developing countries and EITs. A third sub-category was created in 2001, when formal recognition was given to Turkey's "special circumstances" as an Annex I Party.

Non-Annex I Parties

All remaining countries are grouped together as "non-Annex I Parties." The only sub-category within this group consists of the least developed countries, which are granted special assistance and leeway in the submission of their national reports.[3] These countries are not listed in the Convention, the assumption being that the list maintained by the United Nations General Assembly, currently amounting to 49 countries, will be used. The Convention also lists types of developing countries particularly *vulnerable* to climate change or to the negative impacts of mitigation measures and singles these out for special assistance.[4] This category, however, is so general that almost any developing country could argue that it falls under these terms.

The 145 or so non-Annex I Parties consist mostly, but not exclusively, of members of the Group of 77 (G-77), the developing country negotiating group formed in 1964 that was originally composed of 77 countries but now comprises 133 members and is active throughout the U.N. system (China is not part of the G-77, but usually is allied with it). The "G-77 and China" covers a wide spectrum of countries at diverse levels of development and with differing interests relative to climate change, from the small island states, which are seriously threatened by sea-level rise, to the oil-exporting countries, which fear that GHG emission cuts will damage their economies. Despite the diverse interests of its members, the G-77 attributes great importance to maintaining unity, for historical reasons and for accentuating its bargaining strength vis-à-vis the industrialized world. Its individual members are therefore reluctant to argue against the G-77 position, not least because they may require G-77 support on other issues, including those in the broader international arena. Like any group, the G-77 has its more influential members. These, typically the highest aggregate or per capita developing country emitters, tend to be the most wary about taking on formal emission controls.

In addition to the members of the G-77 and China, non-Annex I Parties include the Central Asian countries of the former Soviet Union and several Central and Eastern European states—such as Albania, Armenia, Georgia, and Uzbekistan—along with the new OECD entrants Mexico and South Korea, and a few others, such as Israel.

The Convention and the Kyoto Protocol typically make reference to "developing countries" rather than "non-Annex I Parties." The term "developing country," however, is not defined in the Convention, nor does an official definition exist in the U.N. system. This leads to some uncertainty over the status of countries that are non-Annex I Parties yet do not deem themselves to be developing countries, such as the countries of Central Asia and Central and Eastern Europe, many of which consider themselves to be EITs.

Emission Controls

Emission caps for Annex I Parties

Under both the Convention and the Kyoto Protocol, only Annex I Parties are subject to emission caps. The Convention requires Annex I Parties to take policies and measures "with the aim of returning" their emissions of carbon dioxide and other GHGs to their 1990 levels by 2000.[5] The loose phrasing renders this an aspirational goal, rather than a legally binding commitment. The Kyoto Protocol, however, strengthens the Convention by setting individual, legally binding[6] caps on the emissions of Annex I Parties. Each Annex I Party must reduce its emissions or, in some cases, limit its emissions growth from 1990 levels by the 2008–2012 commitment period (the EITs may apply to use a different base year). The individual targets of the Annex I Parties are listed in the Protocol's Annex B and amount to a collective environmental goal of cutting total Annex I Party emissions by "at least 5 percent" below 1990 levels. The European Union's (EU's) 15 members, which each have an 8 percent reduction target listed in Annex B, have redistributed that target among themselves under a procedure known informally as the "bubble" (UNFCCC 1997a, Article 4). The bubble procedure, which is also allowed under the Convention,[7] is open to any group of Parties that wants to meet its targets jointly.[8]

The emission caps of Annex I Parties were assigned in Kyoto through a process of political negotiation. A variety of objective criteria were proposed, including emissions per capita, emissions per unit of gross domestic product (GDP), GDP per capita, and projected population growth. However, negotiators failed to agree on which criteria to use, with most countries supporting whichever would grant them a more lenient target.

The targets cover emissions of six main GHGs: carbon dioxide, methane, nitrous oxide, hydrofluorocarbons (HFCs), perfluorocarbons (PFCs) and sulfur hexafluoride (SF_6). To help meet emission targets, countries may incorporate limited "sink" activities that absorb GHGs in the land use, land use change, and forestry sector.[9] The precise rules governing the use of sinks under the Kyoto system, including which activities should be covered and what the limits (if any) should be, were hotly negotiated. The resulting deal, agreed to as part of the Marrakesh Accords, has been widely interpreted as a renegotiation of the original Kyoto targets, granting certain Parties—notably Canada, Japan, and the Russian Federation—a considerable number of additional credits.

Box 2.2. Key Features of the Kyoto Protocol

Legally binding emission caps for Annex I Parties. Major characteristics:

- Fixed caps on emission levels, in some cases allowing limited emissions growth and in most requiring reductions
- Set by negotiation, not objective criteria
- Historical base year of 1990, with some flexibility for economies in transition
- Five-year commitment period from 2008 to 2012
- Emissions of six greenhouse gases, plus carbon dioxide absorptions from certain land use, land-use change, and forestry activities. Individual cap on forest management credits.

Flexibility mechanisms include international emissions trading, joint implementation, and the Clean Development Mechanism.

Accountability mechanisms include the measurement, reporting, and review of commitments. The Protocol also includes procedures and mechanisms for dealing with potential cases of non-compliance, as well as mandatory consequences for Annex I countries found to be in violation of certain commitments.

Key features of the Kyoto emission caps, as well as other features of the Protocol, are summarized in Box 2.2.

Involvement of non-Annex I Parties

While only Annex I Parties are subject to emission targets, non-Annex I Parties are also deeply involved in the climate change regime. In addition to hosting projects under the Clean Development Mechanism (see below), non-Annex I Parties have a set of general obligations to address climate change under both the Convention and the Protocol, including commitments to implement and regularly update climate change mitigation programs, promote climate-friendly technological development, and report on emissions and climate policy.[10]

The Convention states that the extent to which developing countries will effectively implement their commitments *depends on the receipt of assistance from the industrialized countries* (emphasis added; UNFCCC 1992, Article 4.7). Annex II Parties must therefore cover the "agreed full incremental costs" of implementing the general commitments of developing countries, along with the "agreed full costs" those countries incur in preparing their reports.[11] Financial assistance from Annex II Parties is mainly channeled through the Convention's financial mechanism, operated by the Global Environment Facility.

Flexibility Mechanisms

The Kyoto Protocol includes three flexibility mechanisms designed to help Annex I Parties meet their targets as cost-effectively as possible. Under the first, *emissions trading*,[12] a Party finding it relatively easy to meet its target may sell surplus emission credits to another Annex I Party finding it more difficult or expensive to stay under its own cap. It must, however, maintain a defined number of emission credits known as a "commitment period reserve" that cannot be traded to minimize the danger that it would then be unable to meet its own target. Through the second mechanism, *joint implementation*,[13] an Annex I Party may fund a specific project that reduces emissions (or increases the uptake of GHGs in the land use, land-use change, and forestry sector) in another Annex I Party, and credit those reductions against its own target.

In practice, transactions under both emissions trading and joint implementation are most likely to take place between the OECD members of Annex I as buyers and the EITs as sellers, given that opportunities to reduce emissions are generally cheaper and more plentiful in EITs. Some EITs—notably the Russian Federation and Ukraine—were granted what many consider rather generous emission targets under the Kyoto Protocol. This has prompted concerns that a large supply of free emission credits that do not result from any mitigation action (so-called hot air) could undermine the environmental goal of the Kyoto Protocol.

The third flexibility mechanism is the *Clean Development Mechanism*[14] (CDM), which works in a similar way to joint implementation, but this time bringing in developing countries as hosts of mitigation projects. The CDM's institutional structure is more complex, including an Executive Board to supervise the system. Its monitoring procedures are also more stringent to guard against the generation of fictitious credits, given that developing countries do not have emission targets themselves and often lack the capacity required to accurately monitor their emissions. In addition to helping Annex I Parties meet their emission targets, the CDM aims to promote sustainable development in developing countries.

Accountability Mechanisms

The Kyoto Protocol's accountability mechanisms[15] aim to safeguard the integrity of the Kyoto system by ensuring that emission reductions are genuine, and not over- (or under-) estimated through error or fraud. They build on similar, less extensive obligations under the Convention, which require Annex I Parties to submit GHG emissions data on an annual basis, as well as regular national reports on their climate policies. The annual emissions data and regular national reports are subject to review by independent "expert review teams" with the power to raise any potential compliance problems with the Protocol's Compliance Committee. A complex accounting procedure,[16] including registry systems at both the national and international levels, will also record transactions in emission credits and allowances. The right of Annex I Parties to fully participate in the three flexibility mechanisms depends on their compliance with these accountability mechanisms.

A *Compliance Committee*[17] will review cases of suspected non-compliance and, if non-compliance is proven, impose penalties through its enforcement branch. Annex I Parties failing to meet their emission targets, for example, must make up the difference plus 30 percent in the next commitment period, must prepare a "compliance action plan," and are banned from selling under emissions trading. The compliance system also includes a facilitative branch and early warning system to help Parties before they fall into non-compliance.

Protection Against Negative Impacts

The climate change regime responds to developing countries' fears concerning their vulnerability, both to the adverse effects of climate change itself and to the possible negative repercussions for their economies of climate policies in the industrialized world (e.g., through reduced demand for oil). To help developing countries adapt to climate change, a levy will be imposed on CDM transactions and the proceeds channeled to a newly created *adaptation fund*, which operates under the Kyoto Protocol.[18] Two other new funds were also set up under the Convention: a *special climate change fund* will finance a variety of adaptation and mitigation projects, including economic diversification for countries heavily dependent on fossil fuel income, while a *least developed countries fund* will finance national adaptation programs of action in least developed countries.[19] Several industrialized countries[20] have already pledged to donate $410 million a year in new money to these funds by 2005.

Institutions and Procedures

Debate and decision-making under the climate change regime continues through the institutions and procedures established by the Convention, which will also be used by the Kyoto Protocol.

The Conference of the Parties (COP) to the Convention, the main body that takes decisions on Convention matters, meets annually, bringing together all the Parties to the Convention. Once the Protocol enters into force, the COP will serve as the Protocol's "meeting of the Parties" (known as the COP/MOP), but only Parties to the Protocol will have decision-making power on Protocol issues. The Convention's two subsidiary bodies—the Subsidiary Body for Scientific and Technological Advice (SBSTA) and the Subsidiary Body for Implementation (SBI)—which usually meet twice a year to carry out more technical work, will also serve the Protocol.

A particular feature of the climate change regime's procedures is that voting rules have never been adopted due to disagreement over what majority should be needed to take decisions. Voting is therefore possible only in cases in which the Convention—and, when it comes into force, the Kyoto Protocol—specifically provides for it, such as the adoption of amendments to the treaties. Even where voting is allowed, Parties are deeply reluctant to go to a vote; governments lay great store on achieving global consensus to move forward on global environmental issues such as climate change, and are loath to openly single out "winners" and "losers" through voting. Decisions in the climate change regime, as with most environmental treaties, are therefore invariably taken by consensus. The overall effect of such a consensus decision-making rule is to give greater power to small minorities, which can threaten to veto the agreement of the majority.

II. Procedures for Further Developing the Climate Change Regime

It was always clear that the Kyoto Protocol, and its emission caps for Annex I Parties for 2008–12, could not provide the definitive solution to climate change. The Kyoto Protocol was intended to serve as a dynamic instrument for long-term climate policy, which could be adapted as Annex I Parties took on stronger targets and non-Annex I Parties assumed formal emission controls.

Two possible pathways—not mutually exclusive—are set out in the climate change regime for further developing its commitments: (1) the accession of individual non-Annex I Parties to Annex I status and Kyoto-type emission caps and (2) the launch of a comprehensive new negotiating round, whose outcome could encompass any of the options for emission controls discussed in this volume. Each of these pathways is discussed below.

Individual Accession

Why would an individual developing country seek to join Annex I and assume Kyoto-type emission caps? The principle incentive—aside from political pressure or a sense of international responsibility—is to enable the country to participate in emissions trading and thereby potentially achieve economic gains. Indeed, this was the main motivation behind Kazakhstan's bid to accede to Annex I.

A developing country wishing to join the Kyoto Protocol's system of emission caps must take two steps. First, it must become an Annex I Party; second, it must have its name, along with an agreed emission target, included in the Protocol's Annex B.

Joining Annex I

A developing country has two possible options to join Annex I for the purposes of the Kyoto Protocol. The first option is for the country to follow the procedure in Article 4.2(g) of the Convention. Under this Article, any non-Annex I Party may notify the Depositary to the Convention—the U.N. Secretary General—that it wants to be bound by the specific commitment of Annex I Parties under the Convention, that is, to take policies and measures with the aim of returning GHG emissions to 1990 levels by 2000. In doing so, the Party would automatically become an Annex I Party *under the Kyoto Protocol* (assuming, of course, that it ratifies the Protocol), as the Protocol's definition of an Annex I Party includes any Party that has made a notification under Article 4.2(g) (UNFCCC 1997a, Article 1.7). The Party would therefore be subject to all the rights and obligations of Annex I Parties under the Protocol, including the right to participate in emissions trading once it has taken on an emission target (subject to its compliance with the accountability mechanisms, as discussed above).[21]

The second option is for the country to add its name to Annex I through an amendment to that annex in the Convention, so that it would become an Annex I Party under the Convention, as well as the Protocol. The two options are not mutually exclusive: A country could seek to amend Annex I in addition to making a notification under Article 4.2(g). The procedures for amending the Convention's annexes are spelled out in Articles 15 and 16 of the Convention.[22] These procedures have already been invoked several times, with mixed results (see Box 2.3).

Inscribing an emission target in Annex B

Even if a non-Annex I Party under the Convention succeeds in becoming an Annex I Party for the purposes of the Protocol, it must take a second step in order to assume an emission cap. Its name, along with an agreed emission target, must be added to Annex B to the Protocol.

This is the stage at which Kazakhstan finds itself in 2002. Having made a declaration under Article 4.2(g), Kazakhstan will be an Annex I Party under the Protocol, but is not listed with an emission target under Annex B. This is, in effect, a "no-man's land." As an Annex I Party, Kazakhstan cannot host CDM projects; however, lacking an emission target, neither can it participate in joint implementation or emissions trading. Belarus and Turkey may find themselves in a similar situation.[23]

Procedures for inscribing a Party's name and emission target under Annex B, however, are more restrictive than for joining Annex I.[24] For a start, there is no notification procedure akin to the Convention's Article 4.2(g) in the Protocol. Instead, the only option open to a new Annex I Party wishing to add its name and a target to Annex B would be to propose a formal amendment to the annex. The procedure for amending Annex B is similar to that for amending the Convention's annexes, except that any Party affected by the amendment must also give its written consent.[25]

While procedures are in place for amending Annex B, the Protocol does not specify how the emission targets for new Annex B entrants should be determined. Given that the existing targets in Annex B were decided through a process of political negotiation, the assumption is that a new Annex I Party wishing to join Annex B would propose a target, and then enter into negotiation with others to establish a figure that would enjoy consensus. Annex B would then be amended to add the Party's name and its agreed emission target.

Prospects for individual accession

Individual accession to Annex I and Annex B may provide a useful entry point to assume emission caps for some non-Annex I Parties that feel ready to do so in the near term, especially those that are not properly considered to be developing countries and are not members of the G-77. These might include Kazakhstan's Central Asian neighbors, which would be granted a

Box 2.3. Amending Annexes I and II

Annex I was first amended at the Third Conference of the Parties (COP 3) in 1997 to add the names of **Croatia, Liechtenstein, Monaco,** and **Slovenia** and to replace the name of Czechoslovakia with those of the **Czech Republic** and **Slovakia.** This uncontroversial decision was made on request of the Parties concerned, and in accordance with Article 4.2(f) of the Convention, which required Annexes I and II to be formally reviewed before December 31, 1998. The Czech Republic, Monaco, Slovakia, and Slovenia all previously had made a submission under Article 4.2(g), while Croatia had declared when it ratified the Convention that it wished to join Annex I.

Also at COP 3, **Turkey** proposed an amendment to Annexes I and II to delete its name from their lists. Industrialized countries, however, did not want to delete a country from the Convention's annexes when their efforts were instead focused on promoting developing country accession. Negotiations dragged on until COP 7 in 2001, at which time agreement was finally reached to delete Turkey from *Annex II*, following acknowledgement that it should not have to provide financial support to non-Annex I Parties, some of which are wealthier than Turkey. Although Annex I was not amended, the COP took a decision formally recognizing that Turkey faces special circumstances, which place it in a different situation than that of other Annex I Parties. How Turkey's special circumstances will be recognized in practice is unclear.

In a third case, **Kazakhstan** proposed to add its name to Annex I through an amendment to the Annex. This proposal met with opposition when it was formally considered at COP 5 in 1999, with some influential developing countries fearing the precedent that would be set if Kazakhstan were allowed to join Annex I. While the new entrants to Annex I at COP 3 were all countries with advanced economies and strong ties to the European Union, Kazakhstan's circumstances are more akin to those of the developing world and the precedent would therefore be more significant. Although Kazakhstan's proposal might have prevailed if put to a vote, the reluctance of Parties to vote under the climate change regime meant this was not politically feasible. After failing to amend Annex I, Kazakhstan took the other route open to it and, in March 2000, exercised its sovereign right to make a notification under Article 4.2(g). COP 7 subsequently recognized that Kazakhstan would become an Annex I Party under the Protocol, but also explicitly noted that it would remain a non-Annex I Party under the Convention.

degree of flexibility as EITs. The recognition of Turkey's special circumstances may also serve as a useful precedent for granting the same status to other prospective Annex I entrants with relatively advanced economies that are not part of the former Soviet bloc.

Moreover, now that the year 2000 is over and, in practice if not in law, the Kyoto Protocol has effectively superseded the Convention's specific commitment to return emissions to their 1990 levels by 2000, the COP could decide to waive this formal obligation for new Annex I entrants. This could help alleviate another concern of non-Annex I Parties, for which returning emissions to 1990 levels is wholly unrealistic.

Accession, however, is not a promising route for the large-scale entry of non-Annex I Parties into an emission control regime. Although, in procedural terms, joining Annex I and Annex B need not be insurmountably complex, it remains a convoluted process, especially in the face of political opposition from other Parties, as illustrated by the case of Kazakhstan. Furthermore, if countries do join during an existing commitment period, this could make it difficult to sustain a collective environmental goal, especially if the new entrant is granted a generous target and could enter the emissions trading market as a large seller of permits. From a political perspective, the prospect of significant numbers of G-77 members breaking ranks with the Group and individually acceding to Annex I is unlikely. The issue of accession to Annex I has acquired considerable negative political baggage over the years. Non-Annex I Parties may be reluctant to join Annex I for the symbolism it implies of joining a "developed country" annex, even if their practical concerns were met.

Overall, therefore, the procedures currently included in the climate change regime for non-Annex I Parties to take on emission controls hold out little promise, by themselves, for the effective future development of the regime. The procedures could be supplemented, however, by the launch of a comprehensive new negotiating round, as discussed below.

Launching a New Negotiating Round

The Convention and the Kyoto Protocol both include clauses that could be invoked to launch a new negotiating round. Both treaties, for example, mandate their respective decision-making bodies, the COP and the COP/MOP, to "periodically examine the obligations of the Parties... in the light of the objective of the Convention."[26] This broad mandate could be raised at any time, provided the decision-making bodies agreed to do so. The Convention and the Protocol also include more specific clauses that serve as hooks to launch new negotiating rounds at a specific time. For instance, it was such a clause—Article 4.2(d) of the Convention requiring COP 1 in 1995 to "review the adequacy" of Annex I Party commitments—that triggered negotiations on what became the Kyoto Protocol.

The Convention hook: Second review of adequacy

The review under Article 4.2(d) of the Convention that led to the negotiations on the Kyoto Protocol was not a one-off event. Article 4.2(d) in fact called for a second review by December 31, 1998, and thereafter at regular intervals. Industrialized and developing countries, however, disagreed over the scope of the second review, generating a deadlock. While industrialized countries wanted to discuss a process for extending emission controls to non-Annex I Parties, developing countries preferred to review the implementation of existing Annex I emission targets. This review process remains "on hold" on the provisional agenda of the COP; despite its tortuous history, it may still provide an opportunity to launch a negotiation on the future of the regime.

A new negotiation round launched under the Convention need not be linked to the Kyoto Protocol and could result in the adoption of an amendment to the Convention[27] or even a new protocol, either for just non-Annex I Parties or for all Parties.

The Kyoto Protocol hooks: Second commitment period negotiations and general review

Article 3.9 of the Kyoto Protocol calls on the COP/MOP to launch negotiations on targets *for Annex I Parties* for the Protocol's second commitment period by 2005. Article 9, in turn, provides for a more general review of the Protocol. This general review is not confined to the commitments of Annex I Parties and therefore could be used to launch a new negotiating round that would also cover non-Annex I emission controls. The first general review of the Protocol is to be carried out by the second session of the COP/MOP. Assuming the Protocol enters into force in 2003, this could be in 2005. These two processes—the launch of negotiations on second commitment period targets for Annex I Parties and a general review of the Protocol—could therefore be held concurrently. Moreover, by 2005, Annex I Parties must have made "demonstrable progress" in meeting their commitments under the Protocol (UNFCCC 1997a, Article 3.2). A review of such "demonstrable progress" might provide the opportunity for Annex I Parties to prove that they have taken the lead required of them in addressing climate change, opening the way to discussing non-Annex I emission controls.[28] The coming together of these various processes points to 2005 or 2006 as a propitious time for launching a new negotiating round.

The most obvious outcome of a new negotiating round launched under the Kyoto Protocol would be for the COP/MOP to adopt an amendment to the Protocol introducing a new set of emission controls—in whatever form—for both non-Annex I and Annex I Parties; as discussed below, quantified emission controls for developing countries would have to be accompanied by deeper emission cuts among the industrialized countries. The amendment procedures for the Kyoto Protocol, which mirror those in the Convention, are set out in its Article 20.[29] Such an amendment could be combined with the adoption of a new annex—Annex C, for example—listing non-Annex I Parties and their new emission controls, just as Annex B does for Annex I Parties.

III. The Challenges of Extending Emission Caps to Developing Countries

The climate change regime is largely silent about its future development, leaving the substantive mandate and eventual outcome of a comprehensive negotiating round completely open. A new negotiation launched through the procedural avenues outlined above could therefore result in agreement on any of the approaches presented in this volume.

Of specific interest for this chapter, however, are the implications of extending the current system of *individual legally binding emission caps* to non-Annex I Parties. The practicalities and challenges associated with "continuing Kyoto" in this way, some of which apply more generally to many of the alternatives outlined in this book, are discussed below.

How Could Developing Country Emission Caps be Negotiated?

Negotiating individual emission caps for 145 diverse non-Annex I Parties poses a daunting challenge. A careful analysis of how such negotiations might work, however, suggests they need not be so difficult. It might be possible, for example, to agree that the 49 least developed countries be exempt from at least the first round of emission caps for developing countries. It might then make sense to give the remaining members of the G-77 space to negotiate among themselves the emission targets of its members, much as the Annex I Parties withdrew into small backrooms to negotiate their own targets in Kyoto, or as the EU negotiated in private the targets of its members under the "bubble." Giving the G-77 space to do this would fulfill the twin goals of taking the diverse national circumstances of developing countries into account and respecting the desire for unity of the Group. It is likely, however, that the larger developing countries with strong trading links to the industrialized world would be compelled to engage in negotiations on their own targets also with the major Annex I Parties.

It is probable that natural groupings would emerge among G-77 members that could share similar targets, on a regional basis, for example. Natural interdependencies may also emerge, with some countries seeking to peg their targets to those of their main economic partners or rivals, perhaps through a formal bubble akin to that of the EU. The South American countries that are members of the MERCOSUR trading bloc,[30] for example, might choose to peg their targets against each other. This is what happened in the Kyoto negotiations; along with Switzerland, most EITs applying to join the EU adopted the same target as the Union. Such groupings and interdependencies would help reduce the complexity of the negotiation.

Negotiations among the other non-Annex I Parties outside the G-77 could take place in a similar fashion. The Central Asian countries would probably wish to take on similar commitments—much as the European EITs on the one hand, and the Russian Federation and the Ukraine on the other, did in Kyoto. For their part, Mexico, Israel, and South Korea are likely to negotiate their targets with their major industrialized economic and political partners.

Alternatively, targets could be negotiated for just a few non-Annex I Parties, with additional groups of developing countries gradually joining the system through new rounds of negotiations over time. While such a staggered approach may appear simpler to negotiate, as fewer targets would be under discussion at any one time, the fact that some G-77 members would have targets and others not would almost inevitably loosen the bonds that hold the G-77 together. Such a challenge to G-77 unity may prove politically unacceptable for the Group.

Adapting the System to Developing Country Needs

While the accommodation of diverse national circumstances is inherent in a system of negotiated emission caps, some further flexibility could also be granted.

Base year

Non-Annex I Parties could, for example, be allowed to select their own historical base year, subject to approval by the COP/MOP. A boundary could be placed around such flexibility by defining a range of years from which Parties could choose.[31] Given that developing country economies are often more volatile than those of industrialized countries, a base *period* may prove more suitable than a single base *year* to help smooth out fluctuations in emissions. Hungary, for example, uses a base period of 1985–87 as part of the flexibility granted to it as an EIT.

Scope of the emission caps

Additional forms of flexibility could also be considered, such as allowing developing countries to select which gases are covered by their targets; and whether, and if so how, to also include the land use, land-use change, and forestry sector. Flexibility may be especially needed concerning reliable data; for instance, such data are often absent for the land use, land-use change, and forestry sector, despite its importance for many developing countries. Moreover, extending the existing rules on the land use, land-use change, and forestry sector to non-Annex I Parties would make for a very complex negotiation, especially as individual caps would need to be set on credits that could be claimed from forest management activities. The process of setting these caps for Annex I Parties was riddled with controversy, and similar difficulties could be expected in negotiating such caps for non-Annex I Parties, especially in the face of data scarcity.

An especially important issue surrounding the land use, land-use change, and forestry sector concerns whether non-Annex I Parties should be permitted to count avoided deforestation against their emission caps, as Annex I Parties currently are. Avoided deforestation was excluded from the scope of the CDM for the first commitment period, partly due to concerns over the difficulty in calculating "real" avoided emissions and the consequent threat to environmental integrity, along with fears over "leakage," whereby trees preserved or planted in one region thanks to a CDM project might simply trigger equivalent tree-felling in another.

The situation would be more complex if sinks, and especially avoided deforestation, were included within the scope of developing-country emission caps. Although the threat of "leakage" would be reduced if non-Annex I Parties had overall emission caps covering sink activities, the threat remains that fictitious credits might be generated, especially in a context of poor emissions data. The impermanence of sink credits is another broader concern surrounding the land use, land-use change, and forestry sector; that is, carbon locked up in trees could be quickly re-released to the atmosphere in the event of a forest fire, for example. These concerns suggest the need for separate rules on the inclusion of the land use, land-use change, and forestry sector in developing-country emission caps or, alternatively, a requirement for countries to demonstrate the existence of sound monitoring and high quality data before this sector can be included.

Accounting, reporting and review

A new system of developing-country emission caps could draw on the Kyoto Protocol's existing accounting, reporting, and review architecture. However, given that non-Annex I Parties are generally at an earlier stage in the development of their emissions estimation and monitoring systems, a significant capacity-building effort would be needed to enable many of them to comply with the highly detailed accounting and reporting obligations currently set forth for Annex I Parties. One option would be to draft less demanding guidelines for non-Annex I Parties under the Protocol, just as there are separate reporting guidelines for these Parties under the Convention. Alternatively, non-Annex I Parties could use the same accounting and reporting guidelines as Annex I Parties, but with some flexibility in their application; the implementation of these guidelines could then be introduced gradually among non-Annex I Parties, depending on their capacities. The aim must be to ensure the maximum possible rigor, in order to maintain confidence in emissions data and therefore in the validity of emission reductions.

Compliance

A related issue concerns the compliance regime under the Kyoto Protocol. Developing countries may call for special consideration by the Compliance Committee in the event of suspected non-compliance; the compliance procedures already make reference to the principle of "common but differentiated responsibilities." It is possible to envisage a sliding scale of such special consideration. At one end of the scale, developing countries could be offered extra assistance from the facilitative branch to help improve emissions data quality or if the early warning system identifies potential compliance problems, but no other special consideration in the event of actual non-compliance. At the other end of the scale, cases of non-compliance by developing countries would be dealt with only through the facilitative branch, without incurring any penalties from the enforcement branch; this would, in effect, render developing country targets legally non-binding.

Legal nature of the caps

At a broader level, it is questionable whether legally binding emission caps are indeed appropriate for developing countries, especially for their first round of quantified targets. Annex I Parties themselves built up experience with legally non-binding pledges in the late 1980s and a soft target under the Convention before they moved on to legally binding caps under the Protocol. A system of voluntary emission caps may be more acceptable to developing countries at this stage, especially given the uncertainty in their emissions data and projections. As discussed in Chapter 1, the term "voluntary" is an ambiguous one, with all commitments in the climate change regime being fundamentally voluntary, in that a sovereign state cannot be forced to sign on to them, and can withdraw its adherence. In this case, however, the assumption is that voluntary *pledges* would be made that would not be considered as binding obligations and, importantly, would not be subject to penalties under the Compliance Committee's enforcement branch. This would have implications, however, for the extent to which developing countries could participate in the flexibility mechanisms; legally binding emission caps are widely seen as fundamental to the emissions trading regime under the Kyoto Protocol.

Flexibility mechanisms

Whether and how to integrate non-Annex I Parties into the emissions trading regime, even if they do take on legally binding emission caps, raises many issues. On the one hand, a primary incentive for non-Annex I Parties to join an emission control system is to participate in emissions trading. On the other hand, their full participation has the potential for generating "tropical hot air." If a large number of non-Annex I Parties, or just a few large emitters, enter the emissions trading market with lenient targets—and therefore a ready surplus to sell—this could lead to a flood of cheap emission credits that could remove any incentive for real climate change mitigation measures among the remainder of the Parties. A related problem surrounds the generally poorer quality emissions data in most non-Annex I Parties, which opens up the possibility that fictitious emission credits might be sold. The combination of potential tropical hot air and poor emissions data means that full participation of non-Annex I Parties in an emissions trading regime, especially one that involves unrestricted trading with Annex I Parties, could undermine the environmental integrity of the Kyoto Protocol.

These concerns can be alleviated to some extent. Although it would not solve the potential problem of overly lenient targets, extending the current eligibility requirements for participating in emissions trading to non-Annex I Parties would mean that they had to meet the existing rigorous accounting and reporting obligations of Annex I Parties before being able to sell emission credits, thus helping to ensure the validity of these credits. Extending eligibility requirements in this way would serve as a clear incentive for non-Annex I Parties to improve their data collection and monitoring systems, although additional financial resources and capacity-building initiatives would be needed to help them make such improvements. Non-Annex I Parties unable to comply with the eligibility criteria for emissions trading might still be able to participate in joint implementation, which has its own supervisory committee for host countries not fully meeting accounting and reporting obligations. Under such a scenario, non-Annex I Parties with emission targets would no longer participate as hosts in the CDM, which would be reserved for any countries exempt from the emission target system.

These are just examples of what could happen. Exactly how non-Annex I Parties with emission targets would be treated under the three mechanisms would be open to negotiation. While such negotiations would undoubtedly seek to minimize the potential for tropical hot air and fictitious credits, the political reality is that the availability of a certain number of cheap emission credits from developing countries may be necessary both to induce Annex I Parties to commit themselves to deeper emission cuts, and to persuade the United States to re-enter the Kyoto regime. Agreeing to the levels of non-Annex I emission targets, and any additional rules for the flexibility mechanisms, so as to fulfill the triple aims of cost-effectiveness, environmental integrity, and attracting both Annex I and non-Annex I Parties into the system, promises to be a difficult balancing act.

Advantages

Familiarity

"Continuing Kyoto" by extending individual caps on absolute emissions to non-Annex I Parties has several advantages. At the most basic level, it has the virtues of simplicity and familiarity. These are at a premium in the climate change regime, where introducing new ideas tends to be a laborious process, due to the cultural, political, and linguistic diversity of the Parties, as well as the limited capacity of many developing countries—some of which have only one person working on climate change—to study, critically analyze, and respond to novel concepts. Continuing with an existing system would bypass the learning process that would inevitably be required for Parties to develop a common understanding of a new concept. In an intergovernmental process involving 186 Parties, complexity can kill even the most intellectually brilliant proposal.

Advance knowledge of environmental benefits

From an environmental perspective, a structure of emission caps has the important benefit of providing an overall goal consisting of the sum of individual targets—such as the collective 5 percent reduction in the Kyoto Protocol—so that the environmental benefits, providing all targets are met, are known when the targets are agreed. This is not the case with emissions intensity targets, for example, whose ultimate environmental impact depends on the subsequent level of GDP growth. Unlike a "top-down" atmospheric concentration target, however, the environmental benefits of a structure of emission caps would not be known when the negotiations are launched.

Full flexibility in implementation

From an economic standpoint, emission caps, as formulated in the Kyoto Protocol, have the advantage of granting full flexibility to Parties as regards their climate policy. Parties can determine which policies they will put in place to meet their targets without restriction, with the only constraint being on the use of carbon sink credits from the land use, land-use change, and forestry sector, and the need to maintain a "commitment period reserve" to participate in emissions trading. Few other proposals for emission control systems give Parties such broad flexibility.

Respect for national circumstances

Extending the current Kyoto system would imply that emission caps for developing countries were set through a bottom-up process of political negotiation, as they were for Annex I Parties in Kyoto, rather than the top-down application of any objective criteria. Allowing developing countries to decide for themselves what target they could commit to, in the context of their complex cultural, economic, social, and political situations, and to negotiate that target with others, would certainly be an obvious and direct way of recognizing their diverse national circumstances.

Obstacles

Political complications

The primary obstacle to extending formal emission controls to developing countries—*be they absolute emission caps or any other form of commitment*—is the general reluctance of these Parties to assume such controls, both because of the low priority given to climate change in the face of more immediate development and local environmental concerns, and the absence of serious leadership from the industrialized countries in tackling their GHG emissions. This is particularly the case among the more powerful G-77 countries. Persuading developing countries as a group to commit to any formal emission controls will be a difficult task until Annex I Parties give clear signs that they have taken the lead required of them in addressing climate change. Although the Annex I Parties as a whole probably returned their emissions to 1990 levels by 2000,[32] this is largely due to the dramatic collapse in emissions in the EITs following the fall of the Soviet Union more than a decade ago. Although emissions have shown a downward trend in some OECD countries (notably Germany and the United Kingdom), emissions in the OECD as a whole rose by 6.6 percent between 1990 and 1999, with some countries (including the United States, Canada, and Australia) experiencing double-digit percentage increases. For the developing countries, many of which also question the extent of effort shown in providing financial assistance and technology transfer, this does not equate with leadership.

Furthermore, in return for assuming formal emission controls, developing countries are likely to insist on two elements: first, deeper emission cuts from the Annex I Parties, and second, a large increase in financial and technological assistance, including for adaptation to climate change impacts. The present obligation on Annex II Parties to fund the "agreed full incremental costs" of developing-country mitigation measures will provide a platform for developing countries to argue that the costs incurred in meeting their new emission caps should also be fully funded. The extent of these costs could be great and hard to delimit. Pressure may also come from the oil-exporting countries for some sort of compensation for any economic loss that they may suffer from this further step toward a less carbon-dependent world. This would make the negotiating round an especially difficult one for the OECD countries; they would be asked to take on larger emission reductions, which many view as costly, and at the same time commit themselves to disbursing increased financial and technological aid.

Although this may not be made explicit, acceptance of emission controls by developing countries will almost certainly be linked to the reentry of the United States into the Kyoto system. It is difficult to imagine the larger non-Annex I Parties, notably Mexico, China, and others with strong trading relationships with the United States, agreeing to emission controls if the United States remains outside the system, even with a large injection of U.S. financial support and technology transfer. The linkage goes two ways, of course, with the United States—at least the current Bush administration—unlikely to ever re-enter the Kyoto regime without formal participation by developing countries. In this respect, a clear disadvantage of "continuing Kyoto" in the current political context is the Bush administration's rejection of absolute emission caps in favor of emissions intensity targets, along with the entrenched negative stance toward the Kyoto Protocol more broadly. "Continuing Kyoto"—even the use of the name "Kyoto" and the political baggage it evokes—may therefore serve to harden the U.S. resolve against the climate change regime rather than encourage its re-entry into an emission control system.

Lack of authoritative emissions data

Lack of authoritative emissions data may prove to be a significant obstacle in setting emission caps for many developing countries. As of mid-2002, more than 60 non-Annex I Parties had not yet submitted their first national reports under the Convention, with the result that no agreed emissions data exist for those countries. Reaching consensus on using any source of data other than officially sanctioned national reports would be difficult. Even these reports often suffer from important gaps in their data sets and, due to differing assumptions and methodologies used, data often cannot be meaningfully compared across countries. Data scarcity, especially in terms of emission projections, would seriously hinder the establishment of meaningful absolute emission caps. This suggests the need for a massive international effort to improve the collection and analysis of emissions data in developing countries *before* the negotiation of their emission controls. The problem of unreliable emission projections, however, is exacerbated by the tendency of developing country economies to be more volatile, an issue discussed further below.

A tendency toward excessively lenient targets?

In regard to in a system of negotiated emission caps, the moral hazard[33] exists that countries would seek the most lenient target possible. It is already assumed that most developing countries will argue for growth caps. The precedent of the Kyoto Protocol, with large growth caps granted to Australia and Iceland and, later, through the EU bubble, to Portugal, Greece, Ireland, and Spain, along with the widely perceived generous targets of the Russian Federation and Ukraine, will not encourage restraint among the non-Annex I Parties. Weaker public interest in environmental issues, along with a scarcity of active environmental nongovernmental organizations, could further reduce the pressure on non-Annex I Parties to take on challenging targets.

While these are real concerns, it would be unwise to assume that negotiated emission caps will trigger a race to the bottom. The target-setting process would not involve self-selection, but *negotiation*, including with those countries most vulnerable to climate change, which are developing countries and G-77 members. Peer pressure from the most vulnerable G-77 members—for example, the small island states and least developed countries—is likely to help shore up the emission targets of the larger developing-country emitters. These most vulnerable countries may not be powerful individually, but they are great in numbers—more than half the G-77—and maintaining a united G-77 front would necessitate taking their interests into account. The experience of the Kyoto negotiations is again valuable in this regard; the United States, Japan, and Canada all took on more stringent targets than they had originally proposed as a result of external pressure and negotiation in Kyoto. Moreover, the COP/MOP would have to approve the new emission caps through the formal process of adopting any amendment or new annex to the Protocol, thus giving any Parties concerned over excessively lenient—and indeed unacceptably strong—targets the chance to veto them.

Are absolute emission caps appropriate for developing countries?

A more fundamental concern is whether absolute emissions caps are appropriate as a form of emission control for developing countries. Absolute caps on emissions are generally viewed, especially by developing countries themselves, as caps on development, while the Convention itself recognizes that development and poverty eradication are the "first and overriding priorities of developing countries" (UNFCCC 1992, Article 4.7). Other options for developing-country emission controls that are more explicitly linked to need (e.g., per capita emission caps, see Chapter 8) or development objectives (e.g., sustainable development policies and measures, see Chapter 3) may provide more politically acceptable solutions.

Moreover, developing country economies tend to fluctuate more than those of the industrialized world because of greater vulnerability to such factors as poor weather conditions or global economic trends. This leads to considerable uncertainty over economic growth rates, and therefore emission projections. A *fixed* emission cap, even spread out over a 5-year commitment period, could therefore prove unmanageable. A country might find that its economy grew much more rapidly than forecasted, thereby overshooting its emission target despite the full implementation of its planned mitigation measures. Conversely, its economy could slump dramatically, generating a large surplus of excess emission credits that could potentially be sold on the emissions trading market, even if no mitigation measures were taken. The combination of uncertainty and fixed emission caps could also encourage non-Annex I Parties to assume optimistic scenarios of economic growth and rapidly rising emissions when negotiating their caps, thus increasing the danger that overly generous targets will be set. The converse, that uncertainty and fixed emission caps might result in excessively tight targets, could lead to widespread non-compliance and, in turn, loss of confidence in the system.

These flaws in the applicability of emission caps to developing countries can be illustrated with reference to Argentina (see Chapter 6). In 1999, Argentina proposed to take on a voluntary target for the first commitment period, once a procedure for doing so was agreed under the climate change regime. Given the absence of any reliable projections of GDP and therefore of GHG emissions, Argentina chose to set a target for itself based on emissions per unit of GDP (emissions intensity), rather than a cap on the absolute level of its emissions. The economic turmoil Argentina has since encountered points to the benefits of such an approach, and the dangers of setting a fixed cap on emissions, especially one that is legally binding, when meaningful economic and emission projections cannot be made.

Conclusion

An important lesson of this chapter is that the design of the climate change regime is, at least in procedural terms, highly adaptable; the institutions of the climate change regime and the procedures for its further development could allow almost any type of emission control to be built into the Kyoto architecture. Given the effort, resources, and extraordinarily complex negotiations that have gone into designing the current regime, it would make sense for any new emission control system to draw on the existing architecture—including its reporting and review system and, where possible, its flexibility mechanisms—as much as possible.

This chapter has pointed to several challenges associated with extending emission controls to developing countries that would apply to most options presented in this volume. Political obstacles, notably the absence of clear leadership from industrialized countries, along with the paucity of reliable emissions data are particularly important. In terms of "continuing Kyoto" through legally binding caps on absolute emissions, the attraction of this option lies chiefly in its familiarity and full compatibility with the existing Kyoto architecture. There are important concerns, however, over the appropriateness of absolute emission caps for developing countries. These concerns point to the need to expand the horizon of possible non-Annex I emission controls beyond the default option. While the Kyoto architecture should be continued, this need not necessarily be done through legally binding, absolute emission caps.

Notes

1. The Czech Republic, Hungary, Poland, and the Slovak Republic have also joined the OECD since 1992, but they were already included in Annex I as EITs.
2. UNFCCC Article 4.6 and Kyoto Protocol Article 3.6.
3. UNFCCC Articles 4.9 and 12.5 and Kyoto Protocol Articles 2.3 and 3.14, which reference Article 4.9.
4. UNFCCC Article 4.8 and Kyoto Protocol Articles 2.3 and 3.14, which reference Article 4.8.
5. UNFCCC Article 4.2(a) and (b).
6. The use of the word "shall" denotes the caps as legally binding. See Kyoto Protocol Article 3.1.
7. Article 4.2(b), which states that Annex I Parties can aim to return their emissions to 1990 levels by 2000 "individually or jointly."
8. Any group wanting to make use of the "bubble," however, must declare exactly how they will redistribute their target when they ratify the Kyoto Protocol, and this redistribution arrangement is then fixed.
9. Limits, for instance, include restricting credits from forest management that an Annex I Party can apply to its individual cap. See UNFCCC (2002, Decision 11/CP.7).
10. UNFCCC Article 4.1 and 12 and Kyoto Protocol Article 10.

11. UNFCCC Article 4.3 and Kyoto Protocol Article 11.2.
12. Kyoto Protocol Article 17 and UNFCCC 2002, Decision 18/CP.7.
13. Kyoto Protocol Article 6 and UNFCCC 2002, Decision 16/CP.7.
14. Kyoto Protocol Article 12 and UNFCCC 2002, Decision 17/CP.7.
15. Kyoto Protocol Articles 5, 7, and 8 and UNFCCC 2002, Decisions 20-23/CP.7.
16. Kyoto Protocol Article 7.4 and UNFCCC 2002, Decision 19/CP.7.
17. Kyoto Protocol Article 18 and UNFCCC 2002, Decision 24/CP.7.
18. UNFCCC 2002, Decision 10/CP.7.
19. UNFCCC 2002, Decisions 5/CP.7 and 7/CP.7.
20. Canada, the European Union, Iceland, New Zealand, Norway, and Switzerland.
21. The legal peculiarities of the climate change regime, however, mean that Parties making a declaration under Article 4.2(g) remain formally classified as non-Annex I Parties *for the purposes of the Convention* and are not bound by the other Convention obligations for Annex I Parties, such as reporting requirements. This was done to encourage non-Annex I Parties to take on the Convention's legally non-binding emission goal, but without also being obliged to meet often onerous reporting requirements.
22. Any Party to the Convention may propose an amendment, which must be circulated to all Parties at least 6 months before the session of the COP that will consider it. The COP can then adopt the amendment by consensus, or, if this is not possible, it can be adopted by a three-quarters majority of Parties present and voting. The amendment to the Annex then enters into force automatically 90 days after being communicated to the Depositary (except for any Parties that notify the Depositary, in writing, that they do not accept it).
23. As they were not Parties to the Convention when the Protocol was adopted, these two Annex I Parties were not included in Annex B. However, Belarus has since ratified the Convention; Turkey, now that its status has been resolved, is also expected to do so. When they ratify the Protocol and it enters into force, they will also be Annex I Parties without an emission target under Annex B.
24. Although the chairman of the Kyoto Protocol negotiations attempted to simplify the Annex B amendment process, Parties were concerned that, if the process were too easy, their names might be added or their commitments changed without their consent.
25. Kyoto Protocol Articles 20 and 21. In addition, once an amendment has been adopted, it must be accepted through a written notification to the Depositary by three-quarters of the Protocol's Parties before it enters into force. Depending on the number of countries that become Parties to the Protocol, this could be a very high hurdle. A procedural means to overcome this hurdle may be for the COP/MOP to agree to provisionally apply the amendment, pending its formal entry into force.
26. UNFCCC Article 7.2(a) and Kyoto Protocol Article 13.4(b).
27. Procedures for adopting amendments to the Convention are the same as for adopting amendments to the Convention's annexes, except that amendments to the

Convention itself must be ratified by three-quarters of Parties before they enter into force, and they only apply to Parties that have ratified them.

28. Reports by Annex I Parties on their demonstrable progress must be submitted by January 1, 2006, and will likely be considered by the COP/MOP in late 2006.
29. Any Party to the Protocol may propose an amendment. See Note 25.
30. Argentina, Bolivia, Brazil, Chile, Paraguay, and Uruguay.
31. Similar to the possibility under the Protocol of choosing 1990 or 1995 as a baseline for HFCs, PFCs, and SF_6.
32. A full set of data to verify this is not yet available.
33. In economics, the term "moral hazard" refers to the effect of certain types of insurance systems in causing a divergence between the private costs of a particular action and the social costs of that action (Pearce 1986).

3. Sustainable Development Policies and Measures: *Starting From Development to Tackle Climate Change*

Harald Winkler, Randall Spalding-Fecher, Stanford Mwakasonda, and Ogunlade Davidson

Introduction

Climate change is a global problem requiring the cooperation of all countries to be addressed effectively. Emissions from the industrialized North have thus far been greater than from the developing South, but they are growing rapidly in the latter.[1] The principle of "common, but differentiated responsibilities" between industrialized and developing countries is well established in the negotiations. However, cooperation between North and South has been limited in the negotiations under the United Nations Framework Convention on Climate Change (UNFCCC). Climate change is not seen as a priority by developing countries, which are preoccupied by the challenges of meeting basic development needs. As the commitment period beyond the Kyoto targets (2008–12) draws closer, the question of how developing countries might participate in the effort against global warming becomes more urgent.

Participation could take different forms. Participation might range from mandatory requirements, such as quantified emission limitation targets, to pledges to make their development path more sustainable. Dividing a global reduction target among all countries (in a "top-down" manner) is only one possible approach (see Chapter 1).[2] The alternative approach is pledge-based (in a "bottom-up" matter). The pledge could be to quantified emission targets, as in the Kyoto process,[3] or more qualitative in nature. In such an approach, it is clear that countries negotiate in their self-interest, so each tends to propose indicators most beneficial to itself (Grubb et al. 1999). Extending the Kyoto regime globally would involve pledges by developing countries (see Chapter 2).

This chapter outlines and proposes a pledge by developing countries to implement sustainable development policies and measures (SD-PAMs). Development is a key priority for decision-makers in developing countries, and therefore building climate change policy on development priorities would make it attractive to these stakeholders. Starting from development objectives and then describing paths of more sustainable development that also address climate change may be the easiest way for many developing countries to take the first steps in longer-term action on climate change. The approach has a basis in the Climate Convention, which, together with a proposed reporting structure, would provide sufficient stringency for a first step.

We begin by outlining the SD-PAMs approach, including its main features and assumptions. In Section II, we apply this approach to South Africa to illustrate the steps taken in practice. Section III considers how this approach might be extended to other countries and which kinds of countries might find it attractive, particularly compared to other approaches. We then consider the relationship of this approach to the ultimate objective of the UNFCCC in Section IV. The conclusion summarizes the major strengths and weaknesses of the SD-PAMs approach.

I. What Is the SD-PAMs Approach?

SD-PAMs is a pledge-based approach to developing-country participation in mitigating climate change. The approach focuses on implementing policies for sustainable development, rather than setting emission targets. The SD-PAMs approach recognizes as a political reality that concerns with climate change (and, in some cases, even environmental policy more broadly) are marginal for many developing countries, and lower in national priority than economic and development policies.[4] It builds on existing commitments and the right to sustainable development enshrined in the Convention.

SD-PAMs differs from the existing "policies and measures" requirements for industrialized countries, which clearly prioritize measures with "impacts in affecting GHG [greenhouse gas] emissions and removals" (UNFCCC 1999). Instead, SD-PAMs starts with the development objectives and needs of developing countries. Countries begin by examining their development priorities and identifying how these could be achieved more sustainably, either by tightening existing policy or implementing new measures. The next step is to identify synergies between sustainable development and climate change, that is, those SD-PAMs that also result in reductions of GHG emissions. To obtain a realistic picture of the impact of a set of SD-PAMs, those policies and measures that increase GHG emissions also need to be identified.

Starting from Development, Shifting to Sustainability

The SD-PAMs approach suggests that we work backwards from a desired future state of development. Key development objectives typically include poverty eradication, job creation, food security, access to modern energy services, transport, drinking water, education, health services, and land. Development is needed because the number of houses to be built, mouths to be fed, and dwellings to be lit and heated is growing.

Sustainability, for the purposes of this chapter, is taken to mean providing for these basic human needs in a way that can continue over time, result in less damage to the environment, and provide more social benefits and long-term economic development. Sustainable development must be driven by local and national priorities. Although documents such as the United Nations Millennium Declaration (UN 2000) and the New Partnership for Africa's Development (NEPAD 2001) articulate goals at the international and regional levels, each country will have its own set of development priorities. The meaning of sustainable development is shaped by the values of each society, and no single approach is appropriate for all economies (Munasinghe 2001, Sachs 1999, Zhou 2001). One of the strengths of the SD-PAMs approach is that it acknowledges and starts from the premise that development and sustainability are country-specific.

In meeting these basic development needs, different paths are possible, and the aim of SD-PAMs is to shift toward a more sustainable path of development. In describing sustainable paths for meeting development objectives, the hypothesis is that, on balance, GHG emissions will also be reduced relative to a conventional development path (Figure 3.1). Many developing countries are already avoiding emissions through current policy. If countries act early to move to even greater sustainability in their development path, they will start "bending the curve" (see Raskin et al. 1998) of their emission trajectory.

This hypothesis is supported by the latest findings of the International Panel on Climate Change (IPCC 2001c). According to the IPCC, a low-carbon future is "associated with a whole set of policies and actions that go

Figure 3.1.
Theoretical Impact of Sustainable Development Policies and Measures on Trajectory of Greenhouse Gas Emissions

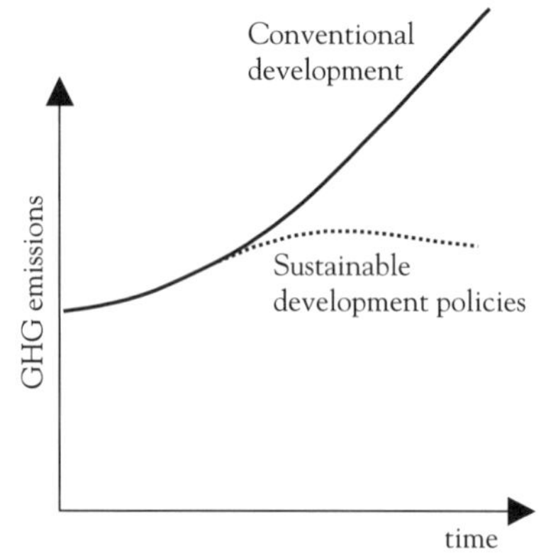

beyond the development of climate policy itself" (Morita and Robinson 2001). Moving toward a sustainable development path could avoid burdensome future mitigation efforts and even have a greater long-term impact on emissions than climate change policies. Thus, the major contribution of SD-PAMs lies not in promoting mitigation effort per se, but in changing the reference scenario of emissions from "conventional" to "sustainable."[5] Likewise, the IPCC also finds that the choice of development path will have a greater impact than climate policy on equity in energy use, suggesting an additional benefit of SD-PAMs (Morita and Robinson 2001, Figure 2.19).

The importance of sustainable development, and its relationship to climate change, has long been recognized in the UNFCCC process. Article 3.4 of the Convention states as a principle that:

> *Parties have a right to, and should promote, sustainable development. Policies and measures* to protect the climate system against human-induced change should be appropriate to the specific condition of each Party and should be integrated with national development programmes, taking into account that economic development is essential for adopting measures to address climate change. (UNFCCC 1992, Article 3.4., emphasis added)

The negotiations, however, have tended to focus more on emission targets than sustainable development, due in part to the predominance of the interests of Northern countries. The links between sustainable development and climate change have received increasing attention in the recent literature.[6] The IPCC's Working Group III has broadened the analysis of climate change mitigation to the context of "development, equity and sustainability" in its contribution to the Third Assessment report (Banuri and Weyant 2001). The challenge considered in this chapter is to turn the conceptual link between sustainable development and climate change into a workable approach.

Global Frameworks and National Circumstances

Climate change policy can be designed to achieve a certain desirable level of atmospheric concentration of GHGs in order to meet the UNFCCC objective of "the stabilization of greenhouse gas concentrations in the atmosphere at a level that would prevent dangerous anthropogenic interference with the climate system" (UNFCCC 1992, Article 2). Given this objective of the Convention, many "top-down" global schemes "backcast" from an assumed GHG concentration target,[7] and then allocate the necessary reductions accordingly across countries.[8] To be successful, those approaches will need to demonstrate how they address the needs of countries and people who face poverty on a significant scale. Such global schemes work out well mathematically, but may have unacceptable consequences for some developing countries.

The impacts of allocation schemes on developing countries are directly correlated with the structure of their energy economies. Primary energy requirements depend on factors such as level of industrialization, economic structure (e.g., presence of energy-intensive industries), level of motorization (car density), average climate (space heating and cooling demands), and domestic energy endowment (predominantly coal, hydro, etc). These national circumstances vary widely among countries and determine national interests and therefore negotiating positions.

The national character of the SD-PAMs approach avoids a "one-size-fits-all" approach to allocating targets. Instead of "backcasting" from a future climate-policy goal, SD-PAMs starts from a country's future development needs and then identifies the most sustainable path of meeting those needs. Starting from a sustainable development perspective "immediately reveals that countries differ in ways that have dramatic implications for scenario baselines and the range of mitigation options that can be considered" (Banuri and Weyant 2001, 76). The SD-PAMs approach, by design, integrates the national development priorities of the country into its approach to climate change. The SD-PAMs approach would be particularly attractive to countries such as South Africa, for which top-down internationally allocated targets may be difficult to agree to or achieve (Winkler et al. 2001).

As the SD-PAMs approach is national in character, it does not have links to international emissions trading. However, implementation of SD-PAMs that reduce GHG emissions are likely to be good candidates for investment under the Clean Development Mechanism (CDM).[9] The CDM requires that projects reduce emissions and promote the sustainable development objectives of the host country; thus, the CDM has a clear synergy with the SD-PAMs approach. Through the CDM and the tradable emission credits generated, developing countries would have some link to the emerging market for carbon credits. The prospect of a Sector-CDM (see Chapter 4) adds further potential because actions under the SD-PAMs approach would involve broader policies (e.g., changes in prices of energy) that could not currently qualify as CDM projects.

Steps in Applying the SD-PAMs Approach

In practice, a country might undertake *five steps* in considering its commitment to SD-PAMs:

1. Outline future development objectives,[10] where possible quantifying the expected benefits and possible risks. If a long-term vision has been articulated, backcasting to immediate action is possible. Otherwise, the country may outline shorter-term goals.
2. Identify policies and measures that would make the development path more sustainable, primarily for *reasons other than climate change* (e.g., greater social equity and local environmental protection while maintaining or enhancing economic growth). The sustainable development benefits should be quantified as far as possible. These SD-PAMs may be the following:
   a. Existing sustainable development policy that is not fully implemented; or
   b. New policies and / or more stringent measures.
3. Quantify the changes in GHG emissions of particular SD-PAMs, which should be reported in accordance with the Convention or other reporting provisions.
4. Compare the results from steps 2 and 3 to show which SD-PAMs create synergies between sustainable development objectives and climate change policy, and which conflict.
5. Summarize the net impact of a basket of SD-PAMs on development benefits and GHG emissions.

Many developing countries already identify development objectives in step 1 through a National Strategy for Sustainable Development, or Agenda 21 plans. To estimate the difference in emissions with and without SD-PAMs, a projection of baseline emissions will be needed in the second step.[11] The information relating to climate change benefits will be useful in implementing and funding SD-PAMs, as those offering greater GHG emission reductions can potentially attract climate change-related funding. Those with greater sustainable development benefits but no climate benefits need to attract other funding. The next section applies this approach to the situation in South Africa. The scope of this chapter does not allow for a full quantification or costing of either the development objectives or the GHG reductions, but examples are provided.

II. Applying SD-PAMs: South Africa as an Illustrative Example

What will the impact of more sustainable development policies and measures in South Africa be on its GHG emissions? To provide a context for this discussion, some background on South Africa's emissions profile is useful.

Context of South Africa's Emissions Profile

South Africa is a semi-industrialized country with an emissions profile that in some respects is not typical of other developing countries. Key characteristics of its economy and energy sector are not favorable in terms of GHG emissions:

- Among major developing countries, South Africa's emissions intensity is relatively high; in 1999, it emitted 0.96 kg of CO_2 per dollar of GDP, expressed in terms of purchasing power parity (PPP),[12] compared to an average of 0.61 among other non-OECD) countries.[13] Reasons for South Africa's high emissions intensity include reliance on coal resources for electricity production, the comparatively low price of electricity,[14] the production of synthetic liquid fuels, a high proportion of energy-intensive industry and mining, and the inefficient use of energy (Winkler and Mavhungu 2001; Spalding-Fecher 2001). Coal-fired power stations account for 93 percent of South Africa's electricity generation.[15]
- Similarly, emissions per capita are high at 8.22 tons of CO_2 (tCO_2) per capita, four times higher than the non-OECD value of 2.11 tCO_2 and higher than several OECD countries (IEA 2001).
- South Africa's share of historical cumulative emissions (1915–95) is somewhat lower (1.17 percent) than its share of 1999 emissions (1.51

percent), reflecting more recent industrialization than in the North (Winkler et al. 2001).

- While South Africa's GDP per capita[16] lies below the world average ($3,160, compared to the global average of $4,890),[17] this figure hides the gap between black and white, and rich and poor, within the country.

Development Priorities

The first step in the SD-PAMs approach is to identify South Africa's development priorities. South Africa's development objectives focus on growth, job creation, and access to key services (including housing, water, sanitation, transport, telecommunications, energy services, and land reform). An overview of South Africa's development objectives was set out in the African National Congress' Reconstruction and Development Programme (RDP) (ANC 1994). It outlined job creation through public works and meeting a range of basic needs as key priorities. However, a new macroeconomic policy, the Growth, Employment and Redistribution (GEAR) strategy, has superseded the RDP (DTI 1996). As the name suggests, GEAR emphasizes economic growth and jobs, while still seeking to redistribute resources. The policy highlights the financial constraints on achieving development objectives, departing from the greater emphasis on social development objectives in the RDP.

Job creation is perhaps South Africa's most important development objective, and is closely related to economic growth. The RDP envisaged large public works programs, which have not materialized. A key element of the vision of GEAR is "a competitive fast-growing economy which creates sufficient jobs for all work-seekers" (DTI 1996), aiming at 6 percent growth and the creation of 400,000 jobs per year.[18] GEAR argues that growth of 3 percent per annum fails to reverse unemployment.

To achieve economic growth, the government aims to reform the labor market, reach inflation targets between 3 and 6 percent, reduce the deficit, accelerate tariff reduction, tighten monetary policy, and limit increases in private- and public-sector wages. Trade liberalization and the privatization of state-owned enterprises[19] are seen as critical mechanisms to promote competitiveness and achieve growth. Spatial development initiatives give a regional focus to the overall objective of economic growth. These initiatives are based in locations where the government hopes to facilitate industrial development through public-private partnerships, the improvement of infrastructure, the establishment of strategic anchor projects, and the creation of industrial clusters and industrial parks (Davis and Wamukonya 1999).

Key to South Africa's development objectives is access to services that meet basic human needs. For the purpose of illustrating the SD-PAMs approach, this chapter focuses on two areas from those listed above—energy and housing.[20] A more comprehensive analysis would require significant effort by a team familiar with all development sectors. Housing and energy are two sectors in which development objectives and GHG changes have been quantified in previous studies. Energy accounted for 78 percent of South Africa's total GHG emissions in 1994 (Van der Merwe and Scholes 1998); housing is a sector in which large sustainable development benefits can be expected.

Energy development priorities

The major objectives of government policy for the *energy* sector, spelled out in the 1998 Energy White Paper (DME 1998), are the following:

- Increasing access to affordable energy services.
- Improving energy governance.
- Stimulating economic development.
- Managing energy-related environmental impacts.
- Securing energy supply through a diversity of energy sources.

Electrification has been a major means of extending access. The first phase of the National Electrification Programme (1994–99) increased access to electricity from 36 percent in 1993 to 66 percent by 1999.[21] The program was internally funded by Eskom, the South African national utility, at a total cost of about R7 billion (Borchers et al. 2001). In 2000 and 2001, a further 734,000 connections have been made (NER 2000, Mlambo-Ngcuka 2002). The government plans to take direct responsibility for further electrification in a restructured power sector. Provision of energy services is not limited to grid electricity. An off-grid rural concessions program has been launched, aiming to provide a total of 350,000 solar home systems in seven concession areas. Proposals have been made to extend the concept to a package that would also include liquefied petroleum gas (LPG) for cooking and other uses (DME 2001a).

A major change in governance of the energy sector is reform of the electricity industry. The way in which restructuring happens in the electricity sector will have significant impact on delivery of services, as well as the future role of energy efficiency and renewable energy (Winkler and

Mavhungu 2001). Opportunities exist for independent power producers to sell renewable energy, but entry into the market is difficult under the current vertically-integrated monopoly system. Public-benefit energy efficiency is likely to be reduced significantly, since private investors have little incentive to invest in measures that reduce revenue (Clark and Mavhungu 2000, Dubash 2002).

In promoting greater diversity in supply, increasing the percentage of renewable energy in the electricity generation mix is a particular goal. The government strategy aims to generate 5 percent of the national grid-supplied power—including import/export—from renewable technologies, mainly from micro-hydro, biomass-fueled turbines, solar thermal, wind turbines, and photovoltaics.[22] The target may be included in the government policy in the White Paper, soon to be published.

Housing development priorities

Addressing the backlog of *housing* is a South African development priority. Estimates of the backlog of houses vary, with Hendler (2000) estimating the number at 2.6 million houses in 1998 and current newspaper reports suggesting a backlog of between two and three million houses (Majola 2002). Roughly three quarters of the housing backlog is urban, and one quarter is rural (Hendler 2000). To meet this challenge, the aim in 1994 was to build 300,000 new units each year of the initial 5-year RDP (ANC 1994). The government provided a housing subsidy of R17,500 for first-time homeowners. The Department of Housing indicates that there were 945,555 "top structures completed or under construction" between April 1994 and July 2000.[23] The RDP outlined that houses should meet basic standards, providing at minimum "protection from weather, a durable structure and reasonable living space and privacy" and access to services, namely "sanitary facilities, storm-water drainage, household energy supply ... and convenient access to clean water" (ANC 1994).

Shifts to Greater Sustainability

Given South Africa's overall development objectives, its sectoral development priorities, and its emphasis on local community development, a number of further shifts to sustainability are possible (step 2 of the SD-PAMs approach). As illustrated in Figure 3.2, current policy probably lies somewhere between a conventional development path and sustainability. Managing energy-related environmental impacts is already part of policy, for example, and is being implemented through programs to promote en-

Figure 3.2.
Theoretical Impact of Current South Africa Policy on Trajectory of Greenhouse Gas Emissions Relative to Conventional and Sustainable Development Paths

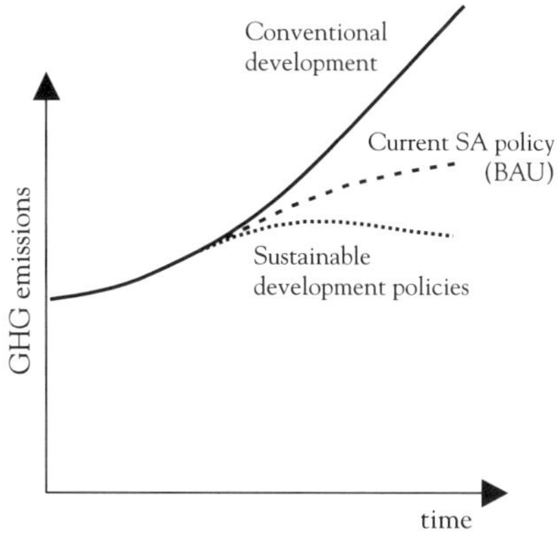

ergy efficiency and renewable energy—even if progress in some areas is still slow (Spalding-Fecher 2001, 10–15). Business as usual (BAU) refers in this chapter to development as stated in current policies, already an improvement on the conventional development path. Emission reductions from BAU therefore do not include GHG changes due to current policy relative to a more conventional development path.

This section outlines possible SD-PAMs for the energy and housing sectors.

SD-PAMs: Electricity

As discussed, the energy sector is a major focus of the government's development objectives. SD-PAMs promoting greater efficiency, increasing the share of cleaner energy, protecting public benefits in liberalizing markets, and providing free electricity can achieve these objectives in more sustainable ways.

i. Efficiency

A national target for greater efficiency in electricity consumption can lead to energy savings, local environmental benefits, and GHG reductions. A recent study (Laitner 2001), using an input-output model of the South African economy, showed that a 5 percent increase in electricity efficiency in 2010 would lead to a net increase of some 39,000 jobs and labor income of about R800 million. The primary reason for the increases is that spending is diverted away from sectors with lower wage and salary multipliers toward construction, finance, and manufacturing, which have higher income multipliers. While not analyzed in detail, a national drive toward energy efficiency of this scale would reduce emissions of carbon dioxide by about 5.5 million tons of CO_2 ($MtCO_2$) in 2010.[24]

End-use energy efficiency by electricity consumers is another measure that saves energy and also reduces GHG emissions. Where energy efficiency reduces overall electricity consumption, it also reduces the overall need for installed capacity.[25] Apart from savings of energy costs, industry often benefits through increased process control and increased productivity. Analysis of one energy efficiency scenario against business as usual by Howells (2000) estimated annual CO_2 reductions of 8 $MtCO_2$ by 2010 and 19 $MtCO_2$ by 2025.[26]

An example of a program to improve end-use energy efficiency is Eskom's Efficient Lighting Initiative, which aims to install 18 million compact fluorescent lights (CFLs) to reduce energy demand in the residential sector (Eskom 2000a). Assuming that the CFLs require only 20 percent of the power for the equivalent incandescent and are used 6 hours per day, Eskom estimates a total energy savings of 4,000 gigawatt-hours (GWh) per year, although this depends on the extent of the "take-back effect."[27] The system average emissions of 0.85 kg CO_2 per kilowatt-hours (kWh) (Eskom 2000b) would imply annual savings of 3.4 $MtCO_2$.[28]

ii. Increasing share of cleaner electricity

The Minister of the Department of Minerals and Energy (DME) has recently re-stated that "renewable energy plays an important role in the energy mix and increases supply security through diversification" (Mlambo-Ngcuka 2002). Achieving this goal has focused so far on developing the Southern African Power Pool, planning increased imports of hydropower, and developing gas markets. Future policy might aim at increasing the share of renewable electricity, which so far has remained in the research, development, and demonstration phase.

A study for the South Africa Country Study on Climate Change (Howells 2000) analyzed the impact on GHG emissions of a cleaner generation mix for bulk energy supply, with a proposed mix consisting of 10 percent nuclear, 10 percent combined-cycle gas turbines, 10 percent imported hydropower, and 1 percent renewables by 2025. Reductions in annual CO_2 emissions against a business-as-usual case were estimated to be 33 $MtCO_2$ by 2010 and 70 $MtCO_2$ by 2025. The costs of the new plants were found to be higher than that of the business-as-usual projection, and the mitigation cost would be about US$2.70 per tCO_2.[29] The emission reductions stem primarily from the increased nuclear, hydropower, and gas capacity, assuming that no GHG emissions are associated with hydropower sources.

A more aggressive policy would be a Renewable Electricity Portfolio Standard. Such a standard might require a basket of options that meets the DME's target of 5 percent of renewable electricity generation by 2010 (Mlambo-Ngcuka 2002). This target may be formalized in a Renewable Energy White Paper, which was under discussion in 2002. The South African Climate Action Network, a group of nongovernmental organizations (NGOs) concerned with climate change, has called for a renewable energy contribution of 10 percent to electricity generation by 2012 (i.e., within 10 years of the World Summit on Sustainable Development) and at least 20 percent by 2020 (SA-CAN 2002).

A first approximation of the impacts of such targets can start with the same baseline emission projection for the bulk energy sector used above. The key assumptions are 2.8 percent annual increase in electricity demand, no climate policy, and new generation capacity, which follow the patterns of the past (Howells 2000). The BAU scenario departs from such conventional development in that it already assumes more advanced and cleaner fossil fuel technologies, an increased share of gas, and more imported hydropower. Assuming that the renewable energy for electricity generation has no emissions and displaces a 2010 generation mix similar to the present (93 percent coal-fired), then the reduction of CO_2 emissions due to 5 percent renewables by 2010 is 10 $MtCO_2$. A shift of 20 percent renewables by 2025 would yield reductions of 57 $MtCO_2$. These reductions are lower than the cleaner generation mix, since that scenario assumed 31 percent of energy supply was low-emissions (i.e., nuclear-gas-hydropower), while this approach proposes increases to only 20 percent. The reductions are significant in the context of 1999 CO_2 emissions from fossil fuel combustion of 346.3 $MtCO_2$ (IEA 2000). The comparative costs of such a portfolio, as well as the impacts on job creation and local economic development, need to be included in future analysis.

iii. Protecting environmental public benefits under restructuring

Greater sustainability in energy governance means maintaining or enhancing public benefits (both environmental and social public goods) in the context of the electric power-sector restructuring process. Determining the GHG impact of such policy interventions is also necessary. Restructuring must provide for new forms of regulation that promote energy efficiency and renewable energy. Distributors may be required to commit a percentage of their total investment to energy efficiency, although the lack of financial viability makes this unlikely in the short term (Winkler and Mavhungu 2001). As of 2002, Eskom was conducting a study on its contribution to sustainability. No estimates of changes in GHG emissions attributable to these policies are available in the literature.

Restructuring also potentially opens access to the grid to independent power producers (IPPs) of renewable energy. Policies and measures required to ensure this happens would include standard contracts for IPPs and non-discriminatory access to the grid. The adoption of a Renewable Energy White Paper with quantified targets for renewable energy generation could set a target.

iv. Providing free electricity—the poverty tariff

The government has committed itself to providing between 20 and 60 kWh of free electricity per month to low-income households. Implementing this "poverty tariff" would provide enough power for poor customers to have access to lighting and entertainment services. If extended to all customers in a broad-based approach,[30] the poverty tariff might at most increase emissions by 0.122 $MtCO_2$, under the assumption that all the free electricity would be additional to existing energy use (UCT 2002). In practice, electricity is likely to displace existing use of paraffin, coal, wood, candles, batteries, and other fuels to some extent. This upper-bound estimate represents 0.03 percent of total GHG emissions, but about 1.6 percent of residential sector emissions in 1994.

SD-PAMs: Housing

How could the delivery of housing be achieved in a more sustainable manner? The DME suggested that "50 percent of all new houses built (including RDP houses) ... incorporate climate conscious solar passive design principles in their construction (thereby eliminating the need for space heating and cooling)" (DME 2001b).

A previous study by the Energy & Development Research Centre (EDRC, an academic research institute based at the University of Cape Town) examined the energy savings, local environmental benefits, and GHG reductions from energy efficiency interventions in low-cost housing (Winkler et al. 2000).[31] The interventions examined focused primarily on improving the energy efficiency in a standard 30-square-meter house[32] and included installing a ceiling, roof insulation, wall insulation, optimizing window size, and adding a partition, as well as a package of all these measures. Interventions in row houses and shacks, as well as lighting and water heating, were also included. The additional cost of these interventions was on the order of R1,000 to R2,000 per household.

The major local sustainable development benefit from these interventions is reduced household expenditure on energy. While small in absolute terms, these savings are significant for low-income households, which devote a relatively large proportion of household expenditure to energy. The interventions also contribute to improved health, because they reduce or eliminate indoor air pollution from burning coal or wood, as well as paraffin fires and poisoning caused by ingestion of paraffin. Energy efficiency may also increase employment if implemented in a labor-intensive program (Irurah 2000).

Energy efficiency reduces energy consumption, and thereby avoids CO_2 emissions from burning fossil fuels, both in homes and in power stations. Avoided emissions were calculated based on the energy savings at the household level using South African emission factors. Interventions that save the most energy for the household (ceilings, wall insulation, solar water heating) also avoid the most emissions.

Taking each intervention and aggregating to the national level, the potential GHG reduction ranges between 0.05 and 0.6 $MtCO_2$ per year, depending on the intervention (Winkler et al. 2000). Although this is a small contribution to potential national emission reductions, the advantage of these mitigation options is their low cost and their significant development benefits.

Changes in GHG Emissions

The third step in the SD-PAMs approach is to consider the changes in GHG emissions resulting from SD-PAMs. These changes in emissions have been outlined for each of the SD-PAMs individually in the previous section and are summarized in Table 3.1. The table also reports the sustainable development benefits and contextualizes the GHG changes, by comparing them with national and, where appropriate, sectoral CO_2 emissions in 1999. On the basis of such information, policymakers could choose the SD-PAMs that best meet multiple objectives.

Table 3.1. Summary of Changes in CO_2 Emissions for Selected Sustainable Development Policies and Measures (SD-PAMs) in South Africa

SD-PAM	Sustainable development benefits	Percentage of CO_2 emissions, 1999	
		Sectoral	*National*
National electricity efficiency improved by 5% (2010)	39,000 additional jobs R800 million additional income	N/a	–2%
End-use energy efficiency (2010)	Energy savings and load management by utility	–5% of CO_2 from electricity	–2%
Share of cleaner electricity increased by 5% by 2010	Reduced local air pollution and fuel costs, increased diversity		–3%
Poverty tariff	Electricity, lighting, and entertainment services from free electricity of 20–60 kWh per household per month for 1.4 million poor households	+1.6% of residential CO_2 emissions	+0.2% (upper bound estimate)
Energy efficiency in low-cost housing	Household energy savings, reduced indoor air pollution, improved health, and increased levels of comfort	–0.6% to –7% of residential CO_2 emissions	–0.01% to –0.2%

Sources: See text on individual SD-PAMs.
Note: The latest estimate of South Africa's total greenhouse gas (GHG) emissions is for 1994; thus, more recent emissions data, covering CO_2 only, are used. CO_2 contributed more than 80 percent of South Africa's total GHG emissions in both the 1990 and 1994 inventories.

A Basket of SD-PAMs?

The SD-PAMs in Table 3.1 are not a comprehensive set; they focus only on two sectors and selected policies. From this initial consideration, however, it appears that most SD-PAMs have more potential for reducing GHG emissions than increasing them. The change in an energy price—that is, the poverty tariff—is the only example of an increase here, yet its impact on overall emissions is small. Since SD-PAMs already include a shift to greater sustainability relative to conventional development,[33] synergies are more likely.

The examples of SD-PAMs from the energy and housing sectors have illustrated some measures with strong sustainable development benefits, some with potential for GHG emission reductions, and some that meet both objectives. Conducting a complete analysis across all sectors would require an interdisciplinary team and significant time and data. Many non-Annex I countries would require assistance in conducting such analyses.

A number of synergies between shifts in sustainable development and GHG reductions are apparent in the energy sector. Energy efficiency is the clearest example, saving on energy costs while reducing GHG emissions. SD-PAMs that promote national electricity efficiency achieve electricity savings, create jobs, add to income, and reduce GHG emissions. A relatively small additional investment in housing for poor communities creates more comfort and reduces household energy costs while cutting emissions from the residential sector.

The poverty tariff provides an example of a conflict between sustainable development and GHG reductions. However, the magnitude of the effect is uncertain, since the degree to which electricity replaces other fuel use is not well known.

Cost has not been explicitly considered in this analysis. In combining SD-PAMs in a basket of measures, some measures that require additional investment have net negative costs over their lifetime. Savings made through energy efficiency could potentially be used to promote a cleaner energy mix. The incremental costs of measures with net costs could be offset against those with net benefits in a basket of SD-PAMs.

Taking the SD-PAMs Approach Further

The last step in the SD-PAMs approach is to consider the overall effect on GHG emissions of a basket of SD-PAMs. Given that this initial study has not covered sectors comprehensively and that some of the SD-PAMs considered here do not have quantified estimates of changes in GHG emissions associated with them, this last step has not been undertaken. Even without this step, the approach identifies areas in which developing countries could act. If, however, this approach is to be linked to a target of global emissions, then this data-intensive step becomes important.

A refinement of the SD-PAMs approach would be to compare stated policy objectives to the country's track record in implementing policies. Projecting this forward (including a gap between stated intentions and actual achievement) might create a more realistic future development scenario. Having illustrated the SD-PAMs approach with the South African example, we consider how this approach could be extended to other developing countries.

III. Extending and Formalizing the Approach

Formalizing the SD-PAMs approach is important not only to monitor whether the commitments are actually implemented but also to challenge perceptions that developing countries are doing nothing on climate change. The materials to formalize the approach can already be found in the Convention and Protocol. Implementing the approach, however, would require some new provisions, including reporting, oversight, and financing.

The Basis of SD-PAMs Commitments

As described in the introduction to this chapter (and in Chapter 1), developing-country participation can take several forms. The Kyoto Protocol sets targets for industrialized countries in the form of binding emission reductions or limits. These commitments are subject to strict monitoring and reporting requirements and mandatory consequences for instances of non-compliance.[34] The SD-PAMs approach suggests a different kind of pledge. As described above, the "commitment" would be to implementing and accelerating national sustainable development plans. Such commitments would initially be voluntary, although they could be made mandatory for at least some developing countries.

The basis for such a commitment is found in the Climate Convention, to which almost all developing countries are signatories. Under Article 4.1(b), all Parties commit themselves to "formulate, implement, publish and regularly update national and, where appropriate, regional programs containing measures to mitigate climate change by addressing anthropogenic emissions by sources and removals by sinks of all greenhouse gases."[35] Using SD-PAMs as a pledge to implement policies for sustainable development would be consistent with Article 10 of the Protocol, which reaffirms existing Convention commitments and aims to "advance the implementation of these commitments in order to achieve sustainable development" (UNFCCC 1997a). This commitment is currently not quantified for developing countries in the same way as for industrialized countries listed in Annex B of the Kyoto Protocol.

Reporting Provisions

While the SD-PAMs commitment would initially be voluntary, a simple reporting system should be established to formalize the commitment of those countries that pledge to implement SD-PAMs. This would require a decision of the Conference of the Parties to establish a registry of SD-PAMs, regular reporting by Parties on their SD-PAMs, and support from the Secretariat for maintaining records of implementation. If voluntary commitments prove successful, a next step would be to make SD-PAMs mandatory for a group of middle-income developing countries. Some developing countries might view this as intergovernmental control over national policymaking, which could present a political obstacle.

This reporting would be similar in spirit to Article 12.4 of the Convention,[36] which says that developing countries may voluntarily propose mitigation projects. The proposed reporting would extend to all SD-PAMs, including those that are not project-based. If countries choose to pledge SD-PAMs, they must report on them and open them for review. In order to assess progress against SD-PAMs pledges, a system of indicators for sustainable development could be adapted from various sources.[37]

Reporting of SD-PAMs could be included in national communications. This would have the advantage that the information would be addressed in the in-depth reviews. However, the process of national communications has become highly politicized, in particular around the provision of technical and financial resources.[38] Given that some developing countries are not submitting their initial national communications, it might be preferable to separate the register of SD-PAMs from this process.

Financing SD-PAMs: Who Pays?

A key barrier to the implementation of SD-PAMs in developing countries is the lack of financial resources. Determining who pays for SD-PAMs is integrally related to the question of formalizing the pledge in the manner suggested above. Countries are unlikely to fulfill pledges unless they have the resources for implementation. Under Article 4.3 of the Convention, developed-country Parties are already committed to paying "full agreed incremental costs" for implementing measures under Article 4.1. If SD-PAMs are adopted under Article 4.1b, the question of payment should in principle be decided already. Where incremental costs are not sufficient, supplementary funding from multilateral institutions, bilateral aid, foreign direct investment, and domestic investment may be needed. For those SD-PAMs with no net implementation costs (e.g., some end-use energy efficiency), only program costs would require funding.[39] Costs of reporting and review should be funded to the "agreed full cost." The commitment to funding is repeated in Article 11 of the Protocol. The challenge is to ensure that funds actually flow.

The sources of funding would differ between those SD-PAMs that have synergies with GHG reduction and those that are neutral or conflict. SD-

PAMs with GHG reduction potential should receive climate change-related funding, including investment through the CDM and Sectoral CDM, climate change funds through the Global Environmental Facility (GEF), and the nascent funds established under the Convention (special climate change fund, least developed country fund) and Protocol (adaptation). Some of these funds would be most suited to projects (CDM), others to enabling activities (GEF) or policy changes (e.g., under Sectoral CDM; see Chapter 4).[40] Providing funding for such projects would be a major incentive for developing countries to take action on climate change. Developing countries could use the SD-PAMs framework to steer financial flows from multiple sources toward climate-friendly sustainable development projects.

SD-PAMs that do not decrease GHG emissions could not draw on climate change funding. They would depend on funding for sustainable development from multilateral institutions, bilateral aid, foreign direct investment, and domestic investments. SD-PAMs also has the potential to harness domestic investment. Further work is needed on the funding of SD-PAMs, especially for implementation.

Which Developing Countries Might Be Particularly Interested in SD-PAMs?

The SD-PAMs approach should be attractive to all developing countries, since its starting point is their own development objectives. The approach should be particularly interesting for developing countries such as South Africa, for which a global allocation provides no surplus credits to sell (and, hence, little incentive to join the system). These are likely to be countries that have already industrialized to a significant extent or, as a result of their particular endowment of energy resources (e.g., large fossil fuel reserves), have used up significant portions of their share of acceptable emissions in a per capita convergence approach (see Chapter 8).

Two ways of indicating which countries might fall into this group would be ranking them by emissions intensity (CO_2 per unit of GDP) and ordering developing countries by ability to pay (GDP per capita). The political criteria to apply to such a grouping would be to include only developing countries and to exclude economies in transition (including the former Soviet republics). Members of the Organization of Petroleum Exporting Countries (OPEC) would rank high in emissions intensity and ability to pay but might nonetheless prefer SD-PAMs pledges to mandatory emission limitation targets. In negotiating developing-country participation, particular attention should be paid to the fact that "global CO_2 mitigation is likely to negatively affect countries that are largely dependent on coal and oil for energy production or export revenues" (Berk et al. 2001, 18). SD-PAMs can offer a "just transition" for communities that would be negatively affected by climate change mitigation.

The approach should also be attractive to least developed countries. The attraction is based on the particularly urgent need for development of least developed countries. A focus on sustainable development would make more sense than any commitment to reductions or limitations of GHG emissions from least developed countries, which are small by international standards.

IV. Relationship to the Climate Convention Objectives

The SD-PAMs approach is a response to climate change starting from development, rather than a commitment to quantified emission limitations targets. While this should be attractive to most developing countries, how does the approach relate to the ultimate objective of the UNFCCC?

Starting from Development

The greatest strength of the SD-PAMs approach is that it starts from a country's development needs and moves toward greater sustainability. Article 2 of the UNFCCC requires that the path to stabilization of concentrations enable "economic development to proceed in a sustainable manner" (UNFCCC 1992). Most developing countries are already committed to doing this. Indeed, they are looking for resources to accelerate this shift.

The approach focuses on the first steps that developing countries might take, rather than offering a one-step solution to the global problem of climate change. Because it matches countries' own priorities, it provides incentives for early action on climate change. Each country would need to consider its own development policies and how those policies could be made more sustainable. The process of formulating development objectives and implementation plans will strengthen coordination between organizations. In this way, the SD-PAMs process will build capacity (politically, technically, financially, and institutionally) in developing countries to tackle policies that reduce emissions. Developing countries can learn by doing by pursuing innovation, development, and transfer of cleaner technologies.

Even within the country, there will be differing views on what the shift to sustainability should entail. Local community benefits—both environmental and developmental—should drive the approach. Tensions between the views of stakeholders from government, business, and civil society are likely to arise. Also, barriers to implementing sustainable practices need to be overcome.

Internationally, a country-specific approach avoids the drawbacks of top-down approaches, which seek to address all countries in the same way and are, invariably, not appropriate to the circumstances of some countries. As long as SD-PAMs can realize the pledge to implementing sustainable development, it has the advantage of starting from each country's unique situation.

Will SD-PAMs Prevent Dangerous Climate Change?

The ultimate objective of the Convention is to prevent dangerous interference with the climate system (UNFCCC 1992, Article 2). This objective is to be achieved in a way that allows ecosystems to adapt, ensures food security, and enables economic development in a sustainable manner. The SD-PAMs approach clearly meets the last condition of achieving the ultimate objective, but does it contribute to stabilization of GHG concentrations?

The answer to this question is indeterminate. The South African example showed that a difficult step is to aggregate the impacts of all the policies and measures. At a global level, the uncertainty is likely to be even larger. It not only requires comprehensive analysis across all development sectors but it is also sensitive to assumptions about the path of future development (which no one knows). This step is critical *if* one wants to compare the result from SD-PAMs to other approaches. It is possible that SD-PAMs would lead to a reduction from business-as-usual emissions but not reduce emissions to "safe" levels if pursued indefinitely.[41] If this were the case, it would undermine the sustainable development of developing countries in particular since they are most vulnerable to the impacts of climate change. Without quantified targets for GHG emission limitations, the SD-PAMs approach cannot guarantee a specific level of global GHG emissions.

On the other hand, striving for a world oriented toward sustainable development will make it easier to meet stringent climate goals, as discussed in Chapter 1, Section I (IPCC 2000a; Berk et al. 2001). If SD-PAMs are really successful, this may even be all that is needed. There is good reason to believe that greater sustainability in development paths will "bend the curve" of emissions. Framing the approach in terms of sustainable development puts incremental decisions in a framework consistent with longer-term targets (see Corfee-Morlot 2002). SD-PAMs can be pursued, even if the net impact on GHG emissions is unknown.

V. Summary

The major strength of the SD-PAMs approach is that it acknowledges each country's unique situation and starts from its own development objectives. The key weakness, from a global climate change perspective, is that it does not guarantee a global reduction in GHG.

The approach may be a useful first step toward developing country participation in climate change mitigation and a learning strategy. If early action on sustainable development leads to effective new markets, technologies, and creative policy solutions, developing countries may later be in a better position to accept other kinds of commitments that quantify emission limitations.

As outlined in Chapter 2, Annex I Parties themselves initially adopted non-binding pledges in the late 1980s and 1990s before accepting quantified and legally binding commitments under the Kyoto Protocol. If SD-PAMs proves robust and successful in reducing GHG emissions, it may be all that is needed in the long term. Moving onto a more sustainable path will build trust for considering other forms of commitments in the future (e.g., third or fourth commitment period). The approach advocates for doing what is possible now and working toward a long-term solution through a series of gradual steps.

Notes

1. IPCC (2001c, 89, but note the caution about use of annual emission for comparison on page 90).
2. Other approaches to developing country commitments are examined in this volume and previous literature (Baumert et al. 1999, Sari 1998).
3. The political process at the Third Conference of the Parties in Kyoto followed a pledge-based approach, rather than a rule-based allocation scheme. Each Annex I country proposed a commitment it might be likely to adopt and, through horse-trading, agreements were struck to reach the final percentage. Characteristics of the industrial and energy economy shaped their national interests which in turn drove their negotiating positions. While arguments were often based on such interests, no systematic quantified analysis of these influences was undertaken. This allowed some industrialized countries to negotiate targets greater than 100 percent of 1990 levels

(Australia 108 percent, Iceland 110 percent, Norway 101 percent) (UNFCCC 1997a). The average global reduction of 5.2 percent reflects no systematic assessment but is simply an average of the voluntary commitments of Annex I countries.

4. See, for example, Mwandosya (2000, 147), Sokona et al. (1999), Berk et al. (2001: 11).
5. In the language of the IPCC emission scenarios, implementing SD-PAMs would help ensure that we are on the path of a more environmentally friendly B1 or B2 world, rather than an A1 world.
6. See, for example, Byrne et al. (1998), Davidson and Nakicenovic (2001), Davidson et al. (2001), ENDA-TM (2001), Munasinghe (2001), and UCS (2001).
7. The IPCC has not defined an atmospheric concentration of greenhouse gases that constitutes "dangerous interference." Different benchmarks are used for illustrative purposes, sometimes the "doubling of CO_2" (about 550 parts per million) or the 450-ppm mark.
8. See, for example, Claussen and McNeilly (1998), Gupta and Bhandari (1999), Redefining Progress (1999), Sijm et al. (2000), Torvanger and Godal (1999).
9. The CDM allows industrialized countries to meet their emission reduction targets by investing in mitigation projects in developing countries, which have no targets. CDM projects must meet the sustainable development objectives of the developing country. Credits for emission reductions are effectively sold to the industrialized country.
10. The default would be to examine development objectives for all sectors. However, some pre-screening of sectors that are deemed most likely to show synergies between sustainable development and climate change could help limit the analysis to a more manageable subset of sectors.
11. Emissions would be reduced in relation to emission projections based on current policy. The biggest problem with doing this relates to high levels of uncertainty about future emissions in developing countries. For SD-PAMs that are project-based, baseline methodologies are being developed through the CDM. For SD-PAMs that require sectoral, multisectoral, or national baselines, further methodological work is needed (see Chapter 4 on sectoral baselines). Politically, such baselines might be seen as similar to a formal commitment, detracting from the voluntary nature of SD-PAMs.
12. Purchasing power parity dollars, using 1990 prices and exchange rates.
13. The previous version of the International Energy Agency (IEA) data—for 1998—showed a more dramatic difference, with South Africa at 1.81 kg CO_2 per dollar of GDP (PPP) compared with a non-OECD average of 0.70 kg CO_2. One reason for the difference may be a change from a base year of 1990 to 1995.
14. Electricity prices in South Africa are low compared with other countries. This does not, however, take into account external costs or the fact that most investments have been paid off. Prices are likely to rise in future.
15. Based on net energy sent out; by installed capacity, the coal share is 89 percent (NER 2000).

16. GDP per capita is not directly part of the emissions profile, but it is a key characteristic shaping a country's ability to pay for mitigation and adaptation.
17. Reported as GNP per capita using exchange rates, based on 1999 dollars, by the World Bank Atlas method (World Bank 2000). South Africa was ranked 86th by this method and 69th when purchasing power parity is used.
18. This objective has not been achieved in past years. The unemployment rate was officially estimated at 25.8 percent for September 2000 (South Africa Reserve Bank 2001), with 11.9 million people employed in February 2000 (Majola 2002).
19. The focus of privatization is on the four big parastatals: Eskom (electricity utility), Transnet (transport), Telkom (telecommunications), and Denel (arms).
20. Several other objectives, for example, providing all citizens with 50 to 60 liters of clean, safe water per person per day; or redistributing 30 percent of land and settling land claims, are not elaborated here. A complete study would need to gather data on all sectors, in particular to complete the fifth step of evaluating the net effect of a basket of SD-PAMs.
21. Access in 1999 remained lower in rural areas (46 percent) than in urban areas (80 percent).
22. The Department of Minerals and Energy produced a draft strategy for Renewable Energy, which is currently being turned into a White Paper (DME 2001b).
23. Department of Housing website, http://www.housing.gov.za/Pages/Indicators/July%202000/wpeD.gif
24. Laitner (2001) gives the figure in units of carbon, that is, 1.5 MtC.
25. In some cases, households may spend energy savings on increasing their consumption, a phenomenon known as the take-back effect. See note 27.
26. A business-as-usual scenario assumed a 2.8 percent increase in demand per year, no climate policy, and new generation capacity following the trends of the past (Howells 2000).
27. One of the major challenges to energy efficiency analysis, especially for the residential sector, is the question of the "take-back," or "rebound," effect: Because energy-efficiency interventions essentially decrease the price of energy services, consumers might spend some of their savings on more of that energy service—so energy consumption may not decline nearly as much as would be predicted on the basis of the technical potential of an intervention. In many developing countries, and particularly in their poorest communities, the level of energy services in poor households is often very low with inadequate lighting, space-heating, and other services, so the rebound effect could be high (Davidson and Sokona (2001), Mehlwana and Quase (1999), Roy (2000), and Simmonds and Mammon (1996)). For the energy-efficient lighting program, the households already have incandescent electric lighting. Given that lighting is often the only electricity service that is affordable for the poor, and that even poor households have several bulbs per household, take-back would be expected to be relatively small (Spalding-Fecher et al., 2002).

28. The baseline against which energy savings from the efficient lighting project are measured would make a significant difference. An earlier study considering different baselines (weighted average or 10th percentile; fuel-specific or sectorwide) found savings ranging between 0.8 and 37 $MtCO_2$ per year. The simple calculation shown here falls toward the low end of this range.
29. R15 per ton of CO_2, converted by the exchange rate for the base year of the data, 1998—R5.53 per dollar (South Africa Reserve Bank 2001)—is the equivalent of US$2.71 per ton of CO_2.
30. The study also considered scenarios in which the poverty tariff is extended only to self-targeted households, resulting in lower incremental emissions.
31. Winkler et al. (2000) is part of a larger research project (Irurah 2000).
32. The standard for RDP houses was initially 30 square meters, but due to strong householder resistance to small units, slightly larger homes (e.g., 42 square meters) have also been built.
33. This argument is strengthened if we consider SD-PAMs against baselines that allow growth. This is explicitly allowed in the CDM rules (UNFCCC 2001, para. 46, p. 37), since the "specific circumstances" of developing countries require development. Analysis of baselines at the project level has suggested that credit should be given for reductions in a situation of suppressed demand (Winkler and Thorne 2002).
34. Monitoring and reporting provisions are outlined in Articles 5, 7, 8, and 18 of the Protocol and have been the subject of detailed negotiations since 1998.
35. UNFCCC (1992, Article 4.1b). The heading of Article 4 is "Commitments."
36. Article 12 deals with national communications, and paragraph 4 reads, "Developing country Parties may, on a voluntary basis, propose projects for financing, including specific technologies, materials, equipment, techniques or practices that would be needed to implement such projects, along with, if possible, an estimate of all incremental costs, of the reductions of emissions and increments of removals of greenhouse gases, as well as an estimate of the consequent benefits."
37. Existing work on indicators for sustainable development in the climate change context includes guidelines and methods developed by the Commission on Sustainable Development (CSD 1995). There is also an ongoing process in the UNFCCC negotiations on "good practices" in policies and measures. For the energy sector, the Helio network has developed and applied sustainable energy indicators (Helio International 2000). A practical method applied to CDM projects (Thorne and La Rovere 1999) could potentially be extended to use at the national level. Chapter 1 of the IPCC's Working Group III Third Assessment Report summarizes the broader debate on sustainable development and climate change, and Chapter 10 focuses on decision analytical frameworks (IPCC 2001c).
38. See the language in UNFCCC (1992, Article 12.7).
39. SD-PAMs would not all be no-regrets or negative-cost options. Indeed, the point of SD-PAMs is to switch the primary focus from emission reductions to sustainable development. This implies assessing cost-effectiveness not only in terms of emissions, but rather in terms of socioeconomic and local environmental benefits.

40. CDM investment is linked to projects and therefore unlikely to fund policy changes, for example, energy policy reforms or industrial strategy. Yet, such policy changes may well be critical to limiting GHG emissions. The Sectoral CDM approach (Chapter 4) would overcome this limitation.
41. Berk et al. (2001, 25) make a similar, but more quantified, argument in relation to the emissions intensity approach: "If the group of countries adopting quantified commitments after the first commitment period would be limited to middle income developing countries, and these countries would initially only take on efficiency improvement targets, and if this would set a precedent for relatively poor, but major developing countries like India and China, CO_2 stabilisation levels of 550 ppmv or lower may be out of reach."

4. EVOLVING TO A SECTOR-BASED CLEAN DEVELOPMENT MECHANISM

Joséluis Samaniego and Christiana Figueres

Introduction

In examining the different options that may be available to shape future climate protection strategies, it is important to recognize the great deal of work that has gone into the current climate regime, based on the United Nations Framework Convention on Climate Change (UNFCCC) and the Kyoto Protocol. Given the fact that the long-standing North-South dynamic will inevitably accompany any further development of the climate regime, it may be advisable to build on existing agreements and current architecture. Chapter 2 explores the default next step of the Protocol: legally binding caps for developing countries. This chapter presents another option for building on the Protocol. Instead of focusing on the commitments assumed by industrialized countries and raising the question of how to integrate developing countries, this chapter focuses on the current avenue for developing country participation, the Clean Development Mechanism (CDM), and explores its possible further evolution. In so doing, this chapter takes the Sustainable Development Policies and Measures (SD-PAMs) approach presented in Chapter 3 and explores its full insertion into the international carbon market through an enhanced CDM.

The Sectoral CDM (S-CDM) approach would maintain some basic elements of the current CDM, but would also allow for the development of CDM projects without pre-established limitations in terms of territorial coverage or enabling instruments (private and public policies and measures). S-CDM "projects" could be sectoral (e.g., electricity, transport, forestry), territorial (entire cities or regions), or a combination of these (such as transport and lighting in a particular city).

This chapter first recounts the evolution of the CDM and its current interpretation. It then presents the envisioned S-CDM, identifying the similarities to the CDM and discussing the contrasting elements. After an examination of the strengths and weaknesses of the S-CDM approach, the chapter profiles a case study of a potential S-CDM project for Mexico City.

I. The Clean Development Mechanism

The CDM is the only flexibility mechanism in the Kyoto Protocol open to developing-country participation. It was established under Article 12 of the Kyoto Protocol and adopted by the Third Conference of the Parties (COP 3) in December 1997. The CDM has a double purpose: to assist developing countries in achieving sustainable development and to help industrialized countries cost-effectively reach the emission reduction commitments they assume under the Kyoto Protocol during the first budget period (2008–12).

Although the CDM was first defined in 1997, the idea is older than the Convention itself.[1] In 1991, Norway introduced the concept of "joint implementation" (JI) during the negotiations that resulted in the UNFCCC. Though termed the same as one of the three flexibility mechanisms later adopted under the Kyoto Protocol, Norway's proposal was broader in definition and constituted a generic term for emissions trading. The concept stemmed from the recognition that the costs of greenhouse gas (GHG) abatement activities vary significantly among countries, and global costs can be reduced if countries form partnerships in their GHG reduction efforts (Dixon 1999). This led to the inclusion of JI in the Climate Convention: "...[P]arties may *implement* such policies and measures *jointly* with other Parties and may assist other Parties in contributing to the achievement of the objective of the Convention[...]" (UNFCCC 1992, Article 4.2(a), emphasis added). Although the Article does not make explicit which countries are meant by "other Parties," the marked difference in abatement costs between industrialized and developing countries soon led to the conclusion that cost-effectiveness would best be served by implementing projects in developing countries or economies in transition.

During the negotiations leading up to COP 1 to the Convention in 1995, representatives of developing countries began to question the value of JI. Some saw it as an attempt by industrialized countries to buy their way out of reduction commitments, particularly if credits for JI projects were to be available before binding targets for domestic emission reductions were in place for the industrialized countries (a step that was not taken until the Kyoto Protocol was adopted). Critics feared that by using

JI projects to achieve low-cost GHG reductions in developing countries, industrialized countries could avoid investments at home and, in this manner, maintain their environmentally unsustainable economies. In addition, some developing countries were concerned that JI projects would exhaust their "cheap" reduction options, so that if emission reductions were to be established for developing countries at a later date, the targets could only be achieved at higher costs (Michaelowa and Dutschke 2000).

Costa Rica was the only developing country that embraced the concept and declared itself available for JI projects as early as 1994. During COP 1, Costa Rica garnered consensus in the G-77 and China group for a compromise proposal. Under a name variation suggested by Malaysia, the "Activities Implemented Jointly" (AIJ) program was established in 1995. A *pilot phase* was introduced to promote "learning by doing" and boost cooperative international efforts. As part of the compromise, no internationally tradable credits would be awarded during the pilot phase, which was to last until the end of the decade.

Between 1995 and 2000, several industrialized countries—in particular, the Scandinavian countries, the Netherlands, Switzerland, and the United States—actively supported the goals and principles of AIJ. They established national AIJ offices and invested in capacity-building activities (Dixon 1999). However, at COP 3 in 1997, the AIJ pilot phase was evaluated and found unsatisfactory. Only a small number of projects had been conducted, due to the lack of incentives in the form of emission reduction credits. Projects were geographically concentrated in Latin America and Eastern Europe and focused mainly on the renewable energy and forestry sectors (Grubb et al. 1999). Neither the distribution nor the mix of project types was considered representative. In addition, transaction costs were very high, and Parties could not come to a consensus on technical issues.

Nevertheless, the concept was not abandoned, but rather was transformed once again. In Kyoto, Brazil suggested the introduction of a penalty system that would subject industrialized countries to a fine if they failed to reach the proposed emission targets. Industrialized countries would have to pay fines in proportion to their degree of non-compliance. The fines would then be channeled into a "Clean Development Fund" and used to support GHG emission-mitigation projects in developing countries and adaptation measures in countries most adversely affected by climate change. Industrialized countries in general, and the United States in particular, were opposed to such a system. The Brazilians were encouraged to change their proposal to a non-punitive concept. The resulting "Clean Development Mechanism" would function as a market-based instrument to channel sustainable development resources to developing countries. Industrialized countries could purchase emission reductions achieved by projects under the CDM to partially meet their reduction commitments. The proposal was backed by G-77 and China, and ultimately approved by the Conference of the Parties under Article 12 of the Protocol.

Designing the CDM was not an easy task. From 1997 to 2000, a wide array of stakeholders around the world developed proposals for the guidelines and modalities of the CDM. As the various proposals were widely discussed and carefully considered, convergence of opinions began to emerge. Agreement on the basic rules and regulations was eventually reached at COP 7 in November 2001. At that time, the first members of the CDM Executive Board were elected and the Board was tasked with writing the detailed rulebook for the CDM. Yet, the decision for a "prompt start" to the Mechanism led to the acceptance of CDM crediting as early as the beginning of the year 2000, and many CDM projects are being prepared as of 2002.

Over the next 10 years, developing countries will be experimenting with the CDM and learning about their mitigation potential. This learning can constitute an important building block for the further development of the climate regime in general, and for the CDM in particular.[2]

II. Sectoral CDM

In looking at ways to strengthen the climate protection regime, this chapter proposes an enhanced CDM as an evolutionary step through which developing countries can increase their participation in the regime.

Characteristics

Under the S-CDM, developing countries would be encouraged to develop regional, sectoral, sub-sectoral, or cross-sectoral projects that may be the result of specific sustainable development policies, measuring the attained reductions, and selling those on the international emission reduction market. Thus, a Sectoral CDM project could be the modernization of the entire cement industry in a country as a result of a government policy, and a cross-sectoral S-CDM project could be achieving a certain efficiency standard in all industrial motors as a result of new standard setting. Table 4.1 provides examples of various types of potential S-CDM projects. Like the SD-PAMs approach in Chapter 3, the S-CDM would involve national or local sustainable development policies. However, in contrast to SD-PAMs,

Table 4.1. Examples of Sectoral Clean Development Mechanism (S-CDM) Projects

Sectoral	Modernization of a country's cement industry
Sub-Sectoral	Conversion of all natural gas-fueled electricity generation plants to combined cycle
Cross-Sectoral	Combination of cleaner transportation and more efficient lighting in one city
Regional	Departure from the business-as-usual emission scenario in one city or other geographic region

the viability of S-CDM—like that of the CDM—is predicated on an explicit link to the international carbon market.

The S-CDM is thus not envisioned as an alternative to the CDM, but rather as a complementary option open to interested countries. S-CDM would build on the current CDM and would have to comply with most of the following existing CDM requirements and design elements:

- **Funding:** As in the CDM, emission reductions achieved through the S-CDM would be sold on the international market to industrialized country entities. In both cases the achievement of emission reductions is financed not by the developing country but rather by offset purchases on the part of an industrialized country entity. Financial institutions that recognize the monetary value of the offsets could fund the implementation. Once attained and certified, the emission reductions could be sold on the international market, becoming part of the project's income flow.
- **Sustainable development:** As in the CDM, the developing country would determine its own sustainable development priorities.
- **Project cycle:** S-CDM projects would undergo the same project cycle as the current CDM. Projects would have to be (1) designed by project participants, (2) approved by the designated national authority, (3) validated by a designated operational entity (third party) and registered by CDM Executive Board, (4) monitored by project participants, and (5) verified and recommended for certification. As a final step, the Executive Board would issue emission reduction certificates (UNFCCC 2002).
- **Additionality:** The Kyoto Protocol establishes that CDM projects may only count emission reductions that are "additional to what otherwise would have occurred in the absence of the certified project activity." As in the CDM, one important goal of an S-CDM project would be the

reduction of emissions or enhancement of absorption relative to a business-as-usual scenario.

- **Verifiability:** As in the CDM, emission reductions or absorptions need to be real, measurable, and verifiable. This demands the use of internationally recognized quantification methodologies and the existence of inventories and reliable projections.
- **No target:** As in the CDM, the S-CDM would operate without legally binding targets for developing countries. In a CDM project, a business-as-usual scenario is defined as the reference case, and emission reductions or absorptions actually generated by the project are measured with respect to that baseline. There is no prior agreement on a target emission level to be achieved, and the developing country can sell all the achieved emission reductions irrespective of its overall emission level. In that sense, the S-CDM does not operate as a "sectoral target" (Philibert and Pershing 2001), but rather as an enhanced CDM.

The S-CDM would build on the current CDM and incorporate many of its characteristics. However, three elements of the current CDM would clearly need to evolve for purposes of the S-CDM (Table 4.2).

Project boundary

While final decisions on what constitutes a project boundary under the CDM have not been made, there is a general assumption that the CDM will only consider single projects (or at most, the bundling of "like" projects). For the time being, the tendency is to prepare and present single projects.

The S-CDM would require a different concept of project boundary. An S-CDM project would have multiple components, not needing boundary definition around each component, but rather around the entire project. The boundary of a sectoral project (e.g., the cement industry) would be easier to determine, as it would include all cement production plants in the country or region. The boundary of a geographically based S-CDM project would be in principle the city or region to which the policies are directed. However, it is entirely possible that not all sectors in a city would be subject to emission reduction policies. For example, the government may choose to include public but not private transportation, or industrial but not residential uses of energy, and so forth. Furthermore, there is the challenge of transboundary emissions (e.g., vehicles traveling in and out of the city). The dispersed nature of mobile sources makes data collection

Table 4.2. Contrasting Elements of the Clean Development Mechanism (CDM) and the Sectoral Clean Development Mechanism (S-CDM)

	CDM	S-CDM
Boundary	Single project	Sector or region
Additionality	Investment in technology upgrade	Policies and measures
Baseline	Project-based	Multiple projects, sectoral or regional

both difficult and expensive (OECD/IEA 2001). Geographically based projects will require further conceptual and technical work on the concept of boundary.

Additionality

The establishment of additionality under the CDM has been the focus of intense debate. Typically, CDM projects introduce a cleaner or more efficient technology or practice. The impetus stems from the project owner's decision to upgrade a specific project with the introduction of state-of-the-art technology. The investment necessary for this GHG upgrade lends the CDM project its additionality.

One of the contentious issues under the CDM is whether a project implemented as a response to a national policy is additional. An example could be the recent switch to natural gas in the public transportation system in New Delhi. A stringent interpretation of additionality would render the investment for the conversion non-additional and thus not eligible for the CDM, reasoning that because the switch was mandated by the government, it would have occurred without CDM intervention.

This interpretation of additionality would not prevail under S-CDM, where such a project might be typical. Just as in the CDM, reduction activities under the S-CDM could be performed by private- or public-sector representatives, but the stimulus to implement the reduction or absorption project would typically arise precisely from a public-sector policy or measure (or even a private sector-led initiative) that pursues both economic development and environmental protection. For the S-CDM to work, sustainable development policies and measures would lead to, and in fact be the basis of, a project's additionality. Under the S-CDM, the expectation is to see many projects reflecting sectoral transformation, such as the above-mentioned New Delhi transport example. The incentives provided by the S-CDM could help trigger these kinds of transformative policies in developing countries sooner rather than later.

Baseline

One of the most difficult issues in the CDM has been, and continues to be, the setting of the baseline—the level of GHG emissions that would have occurred without implementation of the project. During the Kyoto negotiations, developing countries insisted that the CDM be a project-based mechanism, with boundaries and baselines established on an individual project basis. Developing countries feared that multi-project or sectoral baselines could become the backdoor entry to national reduction commitments and were thus determined to keep the CDM clearly on a project-by-project basis. The CDM offers several options for the establishment of a baseline,[3] but all methodologies assume a single specific project.

The S-CDM would have to go beyond single project baselines. The GHG reductions resulting from S-CDM projects that are implemented in response to those policies and measures would have to be measured against an agreed baseline: the emission level or future trend prior to the adoption of the policy or measure within the boundary of the project, be that sectoral, regional, or both.

In some cases, the challenges of baseline setting are exacerbated relative to CDM; in others, the baseline definition is simplified. For example, in the case of a geographically based cross-sectoral project, multiple baselines would probably be necessary, one for each of the components in the project. Here the difficulties of a single project baseline are compounded. On the other hand, in the case of a simple sector-wide project, sector baselines might be easier to establish and monitor. Baselines covering a sector-wide project would also be less prone to leakage; in other words, the project would be able to account for instances in which emission *reductions* from one facility lead to emission *increases* in another.

Advantages

The enhancement of the CDM as an avenue for increased contribution of developing countries to global climate mitigation strategies has many advantages.

Environmental protection

The S-CDM could provide incentives for transforming entire sectors, thus helping to accelerate and deepen the decarbonization of developing country economies. It is unlikely that the current CDM would be able to promote this type of transformation. Current CDM investment is linked to specific projects and therefore is unlikely to promote broad policy changes, such as industrial strategy, more efficient transportation, or cleaner energy mix, as pointed out by Winkler et al. in Chapter 3. Under the S-CDM, the incentive of selling emission reductions at a significant scale may make viable some large, broad-based projects that otherwise would not be undertaken.

As the S-CDM is not based on national targets, it would avoid the moral hazard[4] of developing countries setting lenient targets in order to produce "tropical hot air." The complex procedural structure of the CDM, which would also apply to the S-CDM, is cumbersome and costly but has the advantage of ensuring real, measurable reductions. The S-CDM would help to phase in concrete sector- or region-wide mitigation activities. It is entirely possible that in the short term these activities could deliver more real reductions than if developing countries assumed inflated targets.

Multi-component S-CDM projects could enable GHG emission reductions to take place where costs are very high but the activity is particularly beneficial to national development. Each of the reduction activities included in a multi-component S-CDM project would have different reduction costs. An internal "clearinghouse" mechanism could discover the average reduction cost over the whole project. The single S-CDM project could then place all reductions on the market at market price. Thus, the cheaper reduction components of the project could cross-subsidize the more expensive ones. This kind of mechanism could enable projects that deliver additional, non-climate environmental and social benefits.

North-South equity

One of the key elements of further progress in the climate regime is, indisputably, the acceptance of deeper emission cuts on the part of industrialized countries. The S-CDM reinforces the principle of "common but differentiated responsibilities" by designating industrialized countries as leaders of the mitigation effort. Their future greater reduction commitments are precisely what would create the demand for the S-CDM, a demand that would have to be higher than the current demand for CDM. Thus, the developing countries' increased levels of contribution to climate change mitigation would follow the level of effort of industrialized countries.

In Chapter 2, Depledge points out that under the default option in the Protocol, industrialized countries would likely be asked to take on costly emission reductions, while at the same time being expected to commit increased financial and technological aid for developing countries to meet their increased obligations to the climate regime. The S-CDM presents a win-win option. Industrialized countries are more likely to support a market-based flow of resources to developing countries than increased financial aid. And industrialized countries could assume deeper cuts, as the cost of those cuts would be reduced by the availability of offsets from the S-CDM.

Gradual capacity building

Developing countries need to strengthen their data-gathering and management capabilities. Even if it were politically feasible, it would be difficult in the near term to establish meaningful emission targets for developing countries because of data scarcity and economic uncertainty (see Chapter 5). The S-CDM encourages countries to build up reliable data, sector by sector. Over time, technical capacity, sectoral inventories, and nationwide data can be developed, making any type of future emission controls easier to monitor.

Cost-effectiveness

Currently, the identification, design, negotiation, monitoring, and certification of CDM projects involve high transaction costs. The aggregation or escalation of projects could reduce transaction costs and maximize domestic opportunities for cost-effective reductions. Broadening participation in the market improves the cost-effectiveness of the regime and the market itself.

Adaptation funding

Agreements on the CDM currently stipulate that 2 percent of the proceeds be invested in an Adaptation Fund. The Fund will help defray some of the costs of adaptation in those countries most vulnerable to climate change. If the adaptation share of proceeds in the CDM is held constant for the S-CDM, the higher volume of emission reductions could substantially enhance the funding available to the most at-risk countries.

Compatibility with the Kyoto Protocol

S-CDM is compatible with the present Kyoto Protocol architecture and builds on developing countries' experience. The S-CDM approach could promote a learning process that gradually phases in the participation of developing countries in global climate change mitigation. It could become an important incentive for key developing countries, proportionate to and dependent on increased industrialized country efforts. It could be implemented without major alterations to the structure of the Protocol as it currently stands.

Challenges of Implementation

Despite its advantages, the S-CDM may not be technically feasible or politically viable. By going beyond the CDM, the S-CDM would require an amendment or expansion in the rules governing project boundary, baselines, and additionality in terms that have been discussed. The successful implementation of the S-CDM would have further requirements at both the national and international levels.

At the domestic level, there are two types of challenges to the implementation of the S-CDM.

Technical

Most developing countries are unprepared for the S-CDM, as they have yet to develop the technical capacity needed:

- They must have a functional Designated National Authority with the capability of providing rigorous emissions inventories and projections in order to develop sectoral baselines and monitor aggregated projects.
- Host countries will most likely require an internal "clearinghouse" mechanism, an institutional capacity not common in developing countries.
- Countries must have a reliable GHG accounting system. If S-CDM projects were adopted in a region or sector where an existing CDM individual project is already in operation, the GHG benefits from the single CDM project would have to be excluded from the larger S-CDM project in order to avoid double counting. A clear GHG accounting system is crucial to protecting the credibility of the CDM, and is particularly critical to the S-CDM.

Collaboration

To make some projects viable, domestic institutions not accustomed to collaborating on shared goals would have to develop cooperative strategies. An S-CDM project affecting several sectors in a city—cutting across a wide variety of activities and perhaps even requiring different policy decisions—requires the commitment and political will of a broad set of stakeholders in both the public and private sectors. In addition, the broader the reach of the project, the more important it is to include the participation of civil society in the decision-making process.

At the international level, the S-CDM may face opposition from various negotiation blocs for different reasons.

Developing countries

The CDM has been perceived by some developing countries as weakening the joint effort of industrialized countries to face their climate responsibility.[5] It follows that those same developing countries could perceive the S-CDM as an even greater loophole for industrialized country efforts. After all, the challenges associated with proving additionality in the CDM are not remedied in the S-CDM. The larger scale of the S-CDM raises the stakes of being wrong about the true additionality of a project.

Furthermore, only a few developing countries will have the capacity to design and implement S-CDM projects in the near future. Those countries might command the lion's share of tradable offsets, which otherwise might be distributed among a greater number of countries. This concentration of offsets could exacerbate the inequity between developing countries that receive CDM investment and those that do not, and might cause opposition on the part of those countries that see themselves as disadvantaged by the approach.

Industrialized countries

Industrialized countries might also oppose the S-CDM, since it may have to be concurrent with more stringent emission targets for them. In fact, the viability of the S-CDM may depend on an increased demand from industrialized countries for emission reduction offsets.[6] If industrialized countries remain at the emission limitation levels accepted under the Kyoto Protocol, there would be insufficient demand for a CDM with a supply potential larger than the current one.

Furthermore, this acceptance of deeper cuts on the part of industrialized countries would have to be accompanied by a continuation of the exemption from legally binding targets for developing countries. Industrialized countries may oppose this. If the S-CDM produced a significant amount of emission reductions, industrialized countries would be using their resources to reach their own domestic reduction targets as well as to help developing countries achieve significant reductions. This arrangement flies in the face of the expectation held by some industrialized countries that developing countries, particularly the larger ones, should self-finance their contributions to global climate-change mitigation.

The strengths and weaknesses inherent in the S-CDM are evident in an effort being considered in Mexico City, based on interlinking cross-sectoral GHG mitigation options. If advanced, the effort might well be considered the first S-CDM experiment.

III. S-CDM in Mexico City: A Case Study

Air Pollution and Rising GHG Emissions

Despite some progress achieved in the closing years of the last century, air quality in Mexico City continues to be a major problem affecting the health of a growing population (currently 18 million people). In the late 1990s, the city developed and implemented "PROAIRE," an air-quality improvement program based on cleaner industry, cleaner transportation, urban zoning, and environmental restoration. The first phase of PROAIRE ended in 2000. Its results have been positive on the whole, but much remains to be tackled.

The metropolitan area is a major source of GHG emissions. Annual CO_2 emissions in the Federal District of Mexico, which encompasses much of greater metropolitan Mexico City, amount to about 51 million tons (mt) of CO_2, higher than that of many countries. Projections suggest an increase to 56 mtCO_2 by 2005 and to 63 mtCO_2 by 2010, a growth rate of about 10 percent between 2000 and 2005 and of 23.5 percent over the decade. In all future scenarios—high, medium, and low growth—the metropolitan area expects large increases in the number of inhabitants which, in turn, will raise both energy consumption and CO_2 emissions. The sectoral trends for the next decade show intensified energy use, especially in transportation but also in the industrial and residential sectors.[7]

The capital area represents a large share of Mexico's national totals in both emissions and energy use. The metropolitan area consumes 13 percent of all fossil fuels in Mexico and 17.3 percent of all electricity.[8] The increase in sectoral activity, especially in transportation, is expected to overwhelm the air quality measures undertaken under PROAIRE as well as lead to rising GHG emissions. Despite progress, the policy approach followed so far clearly has to be strengthened. The S-CDM could help provide the financial impetus to strengthen policies and achieve local and global benefits.

Some initial steps are already being considered. As noted above, the first phase of air quality improvements (PROAIRE) was completed in 2000. In preparation for a second phase of the air quality improvement program, the Government of the Federal District (GDF) has also begun to develop a climate change strategy and, in so doing, has shown openness to the S-CDM approach. Under the leadership of Claudia Sheinbaum,[9] the Secretariat of the Environment of the Federal District is well aware of climate change-related issues and of the synergies between GHG mitigation, pollution prevention, and control of urban sprawl. In August 1999, the GDF publicly recognized the need to mitigate climate change on various fronts. In 2000, the new administration established the goal of developing a climate change strategy.[10] Specifically, the GDF has already commissioned studies focusing on specific sectors that could have important emission reduction potentials with local, regional, and global benefits.

The S-CDM Project

From the actions taken thus far, several aspects of the project are already clear:

- First, the potential S-CDM project would have the Federal District as its geographical boundary and would seek to reduce the rapid future emission level rise expected in the District over the next decade.
- Second, these future emission projections would constitute the project's baseline, against which additionality would be assessed. While there would be no fixed emission target, any decrease in expected growth, through the implementation of specific policies and measures across several sectors, would constitute the creditable emission reductions. This benefit would be quantified, monitored, and verified for purposes of the S-CDM.
- Third, the project would capture the positive synergistic effects of policies and measures in support of the S-CDM project undertaken by the GDF. This makes the project different from the simple sum of mitigation actions that might be undertaken by individual sources. Without the S-CDM, a comprehensive, citywide strategy that includes climate change mitigation may be neither feasible nor attractive.

- Fourth, the project would create a local clearinghouse to facilitate reduction of individual efforts within and across sectors.

To advance the possibility of such a project, several concrete steps are being taken. The first important step is developing an inventory of GHGs. The current inventory encompasses the entire urban area within the Valley of México (which is larger than the Federal District). Urban sprawl has caused the metropolitan area to grow beyond the Federal District into the surrounding states of México and Hidalgo. The GDF, however, is responsible for the Federal District only. To have a baseline restricted to the Federal District, the inventory is now being adjusted to identify the share of emissions within the greater metropolitan area that corresponds to the Federal District. Within that geographical boundary, the current emissions from each sector are also being determined.

The second step being implemented is a series of pilot projects. These include an initiative to test fuel cell-powered buses for public transportation (presently in its initial stage) with the aim of introducing this technology more broadly. Other initiatives include the pilot use of solar water heaters, introduction of efficient lighting on a massive scale, testing of electric vehicles, and a carbon sequestration project in the south of the Federal District. However, these efforts are for learning purposes only. They are disjointed and are not achieving their full potential. The GDF is considering an array of measures to integrate the various efforts into a comprehensive mitigation strategy, which could become a coherent plan for a potential S-CDM project.

The potential S-CDM project might encompass simultaneous action in seven sectors within the Federal District: energy efficiency in public and private buildings, industry, new housing, transport, public services, waste management, and reforestation.

1. **Energy efficiency in buildings.** The aim would be to increase the energy efficiency in hotels, hospitals, and other large buildings. Studies show that measures such as insulation, motion-sensor lighting, and efficient water heaters could produce a reduction in energy consumption equivalent to 25.2 megawatts of installed capacity (Government of Mexico 2001).
2. **Industry.** Industrial production in the Federal District contributes to air pollution, but not significantly (relative to transport). However, on days when the health index for air quality reaches emergency levels, industry is forced to shut down. Industry has expressed its willingness

to improve its emission performance and buy the right to continue production processes, through payments for other mitigation efforts, such as reforestation. This willingness to pay opens the possibility of synergizing two sectors in a broad S-CDM project.

3. **New housing.** The GDF is planning to build 10,000 new low-cost homes per year and to remodel 15,000 homes annually. Each home could save 5,000 tons of CO_2 per year through the elimination of liquefied petroleum gas leaks and the installation of solar water heaters. Furthermore, homes are planned with efficient lighting and efficient water pumping. The incremental cost of these new homes has impeded implementation of the planned upgrade.
4. **Transportation.** The Federal District has 105,000 taxis and 21,000 buses. Each taxi emits 75 kilograms of CO_2 per day; each bus emits 230 kilograms of CO_2 per day. The GDF is planning to promote the retirement of old taxis and buses with a subsidy per vehicle replaced. The new vehicles would be more fuel-efficient (in the case of gasoline- or diesel-powered vehicles) and/or feature the use of an alternative fuel, namely natural gas. CO_2 emissions could be lowered by 31 percent in taxis and by 85 percent in buses. In addition, consideration is being given to the introduction of management measures, such as exclusive lanes for public transport and non-motorized vehicles and feeding systems for high-density public transport.
5. **Public services.** General areas of potential emission reductions have been identified, including electricity generation and distribution, water pumping, and wastewater treatment. No specific policies or activities have yet been identified.
6. **Solid waste management.** Both the quantity and the composition of the city's solid waste lend themselves to the possibility of recovering methane emissions for energy generation. Potential is being considered.
7. **Reforestation.** There are a variety of opportunities for reforestation, particularly in the rural southern parts of the Federal District.

These actions incorporate two of the S-CDM challenges: boundary definition and technical capacity to define multi-sectoral baselines. Fortunately, Mexico has already developed much of this capacity, which is not the case in all developing countries. The baseline for the Federal District would need to be grounded in the behavior of each sector and would be contractually binding under the S-CDM project. To verify additionality, actions undertaken in the past would have to be measured and discounted.

Offsets would be only issued *ex post* and in an amount equivalent to the departure from the business-as-usual curve minus the already initiated activities. The reductions resulting from the S-CDM project could also be measured as improved intensity (emissions per unit of local GDP) or as a decrease in the Federal District's rate of emissions growth, depending on the availability of data gathered from additional technical analysis.

Another area for future consideration is the sharing of offsets by project participants. Sharing would need to take place according to predetermined criteria, and the GDF would need to play a central role. The GDF could also use other means (besides offsets) to compensate implementation costs or to provide incentives for emission reductions under the S-CDM project.

The Role of Public Policies and Measures

As a part of an S-CDM project and in pursuit of the implementation of a comprehensive climate change policy, the GDF could establish a range of incentives linked to improved emission performance. Some of these policies and measures could be applicable to specific CDM projects, and others may be generic to the S-CDM approach, but many could gain versatility and reach by being included in the S-CDM approach. Most of them would have impacts on other air pollutants and could be implemented with the double purpose of lowering GHG emissions and improving air quality.

Changes in fuel pricing policies, fiscal incentives (on cars, investments, and so forth), and changes in traffic management policies are a few of the policies and measures possible under an S-CDM project in the Federal District. The GDF Secretariat of Economics has recently considered tax discounts to stimulate desirable environmental behaviors. Another measure that could be considered is establishing a cap on conventional air pollutants or fossil fuel consumption within the Federal District air basin.

The potential S-CDM project in the Federal District harmonizes national and global needs. From the national perspective, urban air quality is a priority. However, previously considered or temporarily implemented policies have not been sufficient to sustain improved air quality and increased health conditions. A well-integrated set of climate change policies, which are important from the global perspective, could catalyze air quality improvement and raise the urban standard of living, if their implementation is at least partially funded by the sale of achieved GHG reductions.

IV. Conclusions

The S-CDM represents a natural evolution of the current climate regime. It would allow developing countries to make serious contributions to the global mitigation efforts without having to take on emission targets. Like the current CDM, S-CDM would offer developing countries the opportunity to pursue GHG-reducing activities with a financial incentive provided by industrialized countries, through either *ex ante* investment or the purchase of resulting tradable offsets. From the perspective of developing countries, other approaches to their increased participation in the climate regime (such as absolute reductions, growth targets, or intensity targets) may appear as a step backward since they call for a reduction-absorption effort funded by domestic resources. From that perspective, S-CDM would maintain the "polluter pays" principle of the CDM, while significantly expanding its scope for emission reductions.

Chapter 2 refers to the "virtues of simplicity and familiarity, which are at a premium in the climate change regime." This points to the potential of the S-CDM, which clearly builds on the learning process of developing countries and obviates the need to introduce a new concept or a new logic into the carefully crafted architecture of the existing regime. Should the architecture be revised in the future, the S-CDM will have served as a bridge toward more demanding approaches and will have given developing countries the opportunity to gain significant mitigation experience.

Some developing countries (Mexico and others that are like-minded) would like to see the S-CDM be a natural enhancement of CDM allowed during the Kyoto Protocol's first commitment period. In principle, this could be possible if the Executive Board does not make decisions that would explicitly impede regional, sectoral, or cross-sectoral projects. If, on the one hand, the COP does not restrict the scale, aggregation, or boundaries of projects in the CDM, the whole idea of proposing a sectoral CDM may be a non-issue. If, on the other hand, the COP decides to explicitly allow CDM projects in which emission reduction offsets result from a set of policies and measures, and not just from a specific technological improvement or infrastructure investment, S-CDM would not be a negotiation point but rather a policy to be fostered by developing countries when designing and operating CDM projects.

However, in the past, when faced with choices on issues such as supplementarity, nuclear energy, and sinks, the COP has shown a clear tendency toward cautious approaches.[11] Therefore, it is likely that the CDM Executive Board will lean toward a narrow (less controversial) definition of project boundary and baseline setting for the first commitment period. Even if the S-CDM is not adopted for the first commitment period, it should still be considered as an option for the future. A number of other future options, including some discussed in this volume (e.g., SD-PAMs) are compatible with the S-CDM. A country that has an emission limitation target could even host an S-CDM project, in the same way that an Annex I country can now host a JI project. Generally, the S-CDM could serve as a valuable transitional mechanism toward future increased participation of developing countries in the global climate change regime.

Notes

1. Portions of this section are adapted from Figueres (2002).
2. The Kyoto Protocol clearly distinguishes the CDM from the other two flexibility mechanisms. Under the Protocol, the term "joint implementation" refers exclusively to the project-based mechanism under which Annex I countries can trade resulting emission reduction units among themselves. These countries also have access to "emissions trading," the buying and selling of emission allowances among themselves. The CDM is the only Kyoto mechanism that involves developing countries.
3. In the CDM, project participants establish the baseline in accordance with internationally approved methodologies. According to the Marrakesh Accords, the baseline can be derived from any of three approaches: existing actual or historical emissions, emissions from a technology that represents an economically attractive investment, or the average emissions of similar project activities undertaken in the previous 5 years under similar circumstances and whose performance is among the top 20 percent of its category.
4. In economics, the term "moral hazard" refers to the effect of certain types of insurance systems in causing a divergence between the private costs of a particular action and the social costs of that action (Pearce 1986).
5. It is important to keep in mind that the reductions or absorptions have benefited the Earth's atmosphere regardless of where they occur. If they are achieved in developing countries, they also contribute to the mitigation of global warming. In this sense, the loophole argument is a subordinate one to the distribution of the effort and not to the outcome of humankind's loading of the atmosphere.
6. There is something of a circular causality in the regime as presently structured. The CDM is the one market instrument that achieves reductions outside of Annex I boundaries, and therefore the only one that, if successful and scalable, could facilitate more decisive efforts among the Annex I countries.
7. Secretariat of the Environment of the Government of the Federal District, personal communication.
8. Secretariat of the Environment of the Government of the Federal District, personal communication.

9. The Secretary of the Environment of Mexico Federal District, previously a member of the Engineering Institute of the National Autonomous University, is a well-known author on energy and climate change issues.
10. Secretariat of the Environment of the Government of the Federal District, personal communication.
11. One clear exception is the fast track of small-scale CDM projects, where the Executive Board is issuing streamlined procedures including sectoral baselines. Experience with these sectoral baselines could serve as a platform for a gradual move toward the S-CDM.

5. Reducing Uncertainty Through Dual-Intensity Targets

Yong-Gun Kim and Kevin A. Baumert

Introduction

Under the Kyoto Protocol, developed countries committed to reduce their emissions from 2008 through 2012 to approximately 5.2 percent below their emissions in 1990. Under this approach, emission constraints of individual countries take the form of *fixed* greenhouse gas (GHG) targets. Such a fixed-target approach may be excessively rigid in the face of shifting economic situations, particularly for developing countries. In unstable developing country economies, reliably forecasting future economic and GHG emission growth is especially difficult. Because of these twin uncertainties, a fixed emission target approach could result in "hot air" in the case of lower-than-expected economic growth or potentially severe constraints on economic development in the case of higher-than-expected economic growth.

This chapter explores two distinct ideas—dynamic targets and dual targets—and their combination, each of which might help reduce these uncertainties. First, *dynamic targets*, where an emission target adjusts in response to another variable, have been proposed for developing countries as a possible future alternative to the Kyoto Protocol's fixed target approach (CCAP 1998, Baumert et al. 1999, Argentine Republic 1999, Philibert and Pershing 2001). Dynamic targets may perform better than fixed targets for economies facing considerable uncertainty, particularly in developing countries. Second, rather than a single target, a *target range* could be established; this approach is called *dual targets*. This chapter examines the viability of *dual-intensity* targets—which combine the ideas behind both dynamic and dual targets. Operating together, dual-intensity targets could further reduce the dangers (e.g., severe reduction burdens or unintended "hot air") stemming from the economic uncertainty in emission target-setting. This approach might also improve the likelihood of reaching a consensus in the climate change negotiations.

Section I of this chapter illustrates the concept and rationale for dynamic targets in general and dual-intensity targets in particular. Section II analyzes economic and emission uncertainties and investigates what these uncertainties imply for dual-intensity targets. It includes a regression analysis illustrating the application of dual-intensity targets for the Republic of Korea (South). Section III discusses several implementation issues, including an analysis of the compatibility of dynamic targets with international emissions trading and other advantages and disadvantages of dynamic targets (and dual-intensity targets specifically).

I. The Dual-Intensity Target Approach

To understand the concept and mechanics of dual-intensity targets, it is necessary to first explore the more general notion of dynamic targets. This section explains several kinds of dynamic targets and elaborates on one kind of dynamic approach—dual-intensity targets. In doing so, this section makes frequent comparisons between *dynamic* targets and *fixed* targets, such as those established in the Kyoto Protocol.

The Concept of Dynamic Targets

The most salient feature of dynamic targets is that they do not establish an absolute cap on a country's allowable emission level. Instead, the allowable emission level for dynamic targets is a function of a predetermined variable; in other words, instead of being fixed, allowable emissions fluctuate in response to some other measure. One can envision the use of numerous variables—including population, previous emissions, and exports. However, economic growth, expressed as gross domestic product (GDP), is the most likely variable because of its substantial influence on a country's overall GHG emissions output. The extent of GDP's influence on overall emissions depends on factors such as the structure of an economy (e.g., predominance of services or industry) and energy mix.

GHG intensity targets

There are at least two kinds of dynamic targets. One is often termed an emission "intensity target" (Baumert et al. 1999). Here, the target itself is expressed not in terms of an absolute measure, such as tons of GHGs, but in terms of an emissions intensity—a ratio between GHG emissions and economic output:

$$\text{Intensity Target, } I = \text{Emissions/GDP}^{\alpha}$$

where I is the emissions intensity target—a constant, expressed in tons of GHGs per unit of GDP. *Emissions* is the country's allowable emission level during the target period. *GDP* is the country's aggregate gross domestic product during that period, and α is a multiplier that determines the manner in which the allowable emission level changes in response to GDP. If α is equal to 1, then the relationship is linear: a 1 percent increase in GDP will increase the allowable emissions by 1 percent (because I is constant). In the case of Argentina (Chapter 6), α was set at 0.5 (i.e., the square root of GDP).

This formula can also be expressed as Emissions = $I \times \text{GDP}^{\alpha}$. Here, plugging the actual GDP value into the equation will yield the allowable emissions amount, because I and α are constants.

Indexed targets

A second kind of dynamic target uses indexing to adjust the allowable emission level (Frankel 1999). Like intensity targets, indexing adjusts the allowable emission level according to changes in GDP. Here, the agreed target (i.e., the allowable emissions) would be accompanied by an assumed annual average rate growth (AAARG) of GDP. Deviations from this assumed rate of GDP growth would trigger adjustments in the allowable emission level. For example, country Z adopts an emission target that limits its emissions to 100 units during a particular period. Z's target assumes that the average rate of GDP growth will be 4 percent annually (i.e., AAARG = 4 percent). If actual GDP growth exceeds 4 percent per year, the target is adjusted upward. An annual rate of GDP growth of, for instance, 6 percent (i.e., 2 percent in excess of the assumed rate) might enable the emission level to increase by 2 percent for every year between the negotiation and compliance dates. Conversely, if actual GDP growth is less than 4 percent per year, the target is adjusted downward.

For indexed targets, the adjustments do not need to be linear, just as, in the case of intensity targets, α does not have to equal 1. For example, GDP growth of 1 percent higher than the AAARG could result in an increase in emissions of 0.75 percent; growth of 1 percent *less* than the AAARG might result in a decrease of 0.50 percent in allowable emissions.

It is important to note that, while "intensity targets" and "indexed targets" may appear different, they are essentially the same. Under each, the

Table 5.1. Analyses and Proposals on Dynamic Targets

Source	Target indicator	Other characteristics
CCAP (1998)	Growth Baseline: Emissions/GDP	"Carbon efficiency" (C/GDP) target between BAU and a no-regrets baseline
Baumert et al. (WRI 1999)	Intensity target: Emissions/GDP	Reduction in intensity relative to BAU, measured from a historical base year
Argentine Republic (1999)	Emissions/$\sqrt{\text{GDP}}$	Reduction in emissions of between 2 and 10 percent relative to BAU (nine scenarios); legally binding if emissions trading is allowed.
Frankel (Brookings 1999)	GDP-indexed target	Target established at the BAU level or lower (approaching a "break-even" level, where gains from trade equal domestic costs).
Philibert (IEA/OECD 2002b)	GDP-indexed target	Possible use of price cap or other measures to enhance flexibility
U.S. Administration (2002)*	Intensity target: Emissions/GDP	Reduction in greenhouse gas intensity by 18 percent (relative to 2002) over 10 years; voluntary agreement
Lutter (2000)	Emission/[(lagged emission)$^{0.5}$×(lagged GDP)$^{0.6}$×(lagged GDP per capita)$^{0.06}$]	

* For U.S. administration, see White House 2002. **Abbreviations**: GDP (gross domestic product), BAU (business as usual).

allowable emission level fluctuates with economic activity. Table 5.1 summarizes the different dynamic targets that have been analyzed or proposed.

Economic uncertainty

Future GHG emission levels are highly uncertain in developing countries (a topic explored in greater detail below). This situation can lead to serious technical difficulties in establishing a future GHG emission limitation using a fixed target. Achieving a specific future level of GHG emissions might be very easy under conditions of low economic growth, industrial stagnation, and population decline. That same GHG goal, however, might be exceedingly difficult to reach if economic growth were instead robust and population were increasing. Thus, fixed GHG goals can entail widely varying levels of effort, depending on underlying socioeconomic conditions (especially GDP growth), which tend to have a powerful influence on emission levels.

This represents a serious problem. Experience suggests that when countries are proposing or evaluating a potential emission target, they are particularly concerned with economic impacts. In other words, countries want to know the impact that a particular emission control target will have on their domestic economy, including the overall costs and benefits, potential job losses and gains, and changes in international competitiveness. If a developing country were to agree to a fixed emission target, potential economic impacts are likely to be highly *uncertain*. Dynamic targets attempt to address this uncertainty by adjusting to economic reality and therefore reducing the economic uncertainty associated with taking a particular target. They allow faster-growing economies more emissions and contracting economies fewer emissions.

Governments are risk-averse with respect to economic considerations, such as growth, jobs, and competitiveness. This is especially the case in developing countries, where climate change is not a priority. If a developing country contemplates a GHG target, it will be important that this target does not unreasonably impinge on its development prospects. Given their risk aversion, developing countries might avoid GHG targets that have the *potential* to adversely affect economic growth, even if that potential is small.

Environmental uncertainty and environmental effectiveness

With dynamic targets, reduced economic uncertainty comes at the expense of environmental certainty. Unlike Kyoto-style fixed targets, dynamic targets do not guarantee any particular environmental outcome, although they will deliver environmental outcomes within a relatively predictable range.

It is important that the reduced upfront environmental certainty of dynamic targets is not equated with weaker environmental outcomes. Dynamic targets could actually facilitate more stringent emission limits, due to the reduced economic uncertainty of such targets, discussed above (Baumert et al. 1999, Philibert 2002a). Given governments' risk aversion, a fixed target could create an incentive for a developing country to settle only on a weak target that ensured no economic harm. Weaker emission limits are a serious drawback of fixed targets, especially given the links between emission targets and international emissions trading. Weak targets for one country (inadvertent or not) can reduce the environmental effectiveness of the entire regime by allowing other countries to purchase and use excess emission allowances that otherwise would not be used. Such excess allowances are often referred to as "hot air." Although hot air can be a political creation (from negotiating emission limits in excess of future needs), it can be enhanced by unexpected declines in economic activity after a fixed target has been negotiated.[1] Overall, dynamic targets do not eliminate the risk of negotiating hot air targets, but they do provide upfront transparency that will at least help Parties identify whether a target is likely to generate hot air.

In addition, dynamic targets could enhance environmental effectiveness by promoting wider participation in the international emission control system. Fixed targets might be simply unacceptable for many developing countries. Given the unsettling choice in the target-setting process—between weak targets (which would do little to help the global environment) and strong targets (which could have deleterious effects on their domestic economies), developing countries might opt for *no* commitment along these lines. Wider participation also supports environmental effectiveness by reducing the incidence of emission "leakage" from countries with emission constraints to those without.

Sustainable development

Another compelling feature of dynamic targets is their compatibility with sustainable development because they are geared toward achieving emission reduction *relative* to economic development rather than achieving absolute reductions in emissions (Baumert et al. 1999). Intensity indicators might better reflect the real climate challenge in developing countries—decoupling economic growth and emissions growth. Philibert (2002b) also states that dynamic targets could be considered most compatible with the environmental strategy adopted by the OECD, which is mainly based on the concept of "de-coupling environmental pressures from economic growth."

The Concept of Dual Targets

As discussed, dynamic targets are considered more appropriate than fixed emission targets with respect to accommodating uncertain economic growth rates, especially in developing countries. However, many difficulties and uncertainties remain in establishing emission targets for developing countries, whose economic growth is highly unpredictable.

Just as with fixed targets, a developing country would have an incentive to overestimate its "business-as-usual" (BAU) emissions intensity (to justify a weaker target) while other, rival negotiating countries would have the opposite incentive. In fact, if a country hypothetically tried to establish a target representing its BAU emission levels, even with a dynamic target there would likely be some degree of either hot air or economic burden. Thus, this section explores the possibility of establishing *dual* targets that, taken together, cover a range of future scenarios. This approach might improve the effectiveness of target setting and make consensus easier to reach.

The concept of dual targets is not new. Philibert and Pershing (2000) proposed to establish two national targets with differing legal characters: one non-binding, the other binding. The binding target would allow a relatively high level of emissions to prevent the risk of undue constraints on economic growth. The non-binding target would be established at a more stringent level, in order to reduce the risk of hot air. This non-binding target would be the "selling target," while the binding target would be a "buying target." Although Philibert and Pershing did not consider the dual-target concept as an option for dynamic targets, it deserves to be extended and generalized to a wider policy design context. Dual targets could be applied to either fixed or dynamic targets. The *dual-intensity* target proposal in this chapter effectively combines the intensity-target approach and dual-target concept.

Combining the Concepts: Dual-Intensity Targets

Under the dual-intensity target approach, two emissions-intensity targets are established for a single country. The two targets have separate purposes. The lower (more stringent) target provides an incentive to reduce emissions: reductions below this target would enable the country to sell emission allowances. The higher (less stringent) target would have a punitive function: Exceeding this target would require the country to purchase excess emission allowances in order to remain in compliance. Thus, the lower target would be the "selling target" and the higher one termed the "compliance target." No penalty applies if the emissions intensity of the country lies between the selling and the compliance targets. That is, there is a "safe zone" in which the country is neither out of compliance nor able to sell allowances through international emissions trading. This approach is illustrated in Figure 5.1 (the "safe zone" is the dark shaded rectangle between the selling and compliance targets). Mathematically, the two formulas would take a form similar to the intensity formula described above:

Selling Target: $Emissions = I_1 \times GDP^{\alpha}$
Compliance Target: $Emissions = I_2 \times GDP^{\alpha}$, $I_1 \le I_2$

Figure 5.1. Graphical Illustration of Dual-Intensity Target

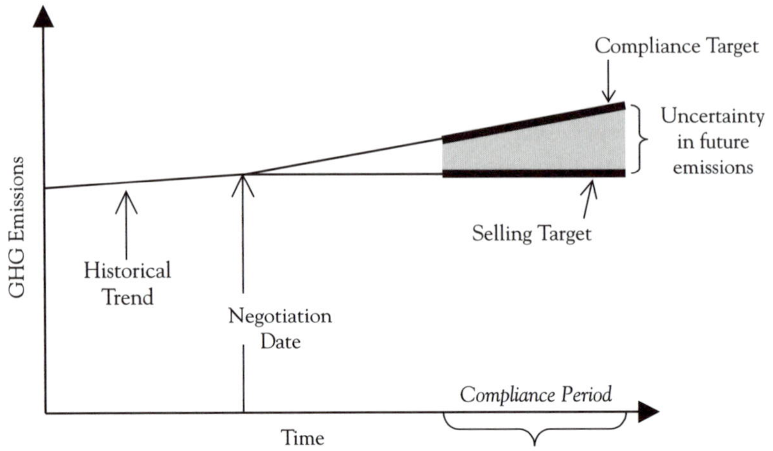

I_1 denotes the lower (selling) intensity target and I_2 the higher (compliance) intensity target. This formulation is general enough to encompass a wide range of alternatives. If we set "$I_1 = I_2$," it is identical to a single intensity target. If we set "$I_2 = \infty$," it implies an incentive-only intensity target where there is no obligation to limit emissions (but also no trading, unless the selling target is reached). Therefore, the concept of dual-intensity targets gives us a general and flexible framework for commitments.

II. Analysis of Uncertainties and Implications of Dual-Intensity Targets

An underlying premise of dual-intensity targets is that uncertainty in future GHG emission levels impairs the process through which emission targets are set. Thus, it is worth illustrating the underlying economic and emission uncertainties in more detail and showing how dual-intensity targets help policymakers manage this uncertainty. This section analyzes historical and projected data in an attempt to derive implications for the performance of dynamic targets, particularly dual-intensity targets. First, the uncertainties in future projections of both emissions and emissions intensities are compared. Also, examples of past longer-term projections are scrutinized to see how those forecasts tend to change significantly over time. Finally, the result of a regression analysis for the case of Korea is presented and a potential application of dual-intensity targets is described.

Uncertainty of Forecasts: Experiences from Past Projections

One can evaluate different indicators by analyzing the reliability of past forecasts and the uncertainty of future projections. Table 5.2 shows CO_2 emissions and intensity (i.e., CO_2/GDP) projections to 2020. These projections, undertaken by the U.S. Energy Information Administration (EIA), include three scenarios: a reference scenario (i.e., BAU), a high GDP-growth case, and a low GDP-growth case. The table shows the range between the high and low emissions and intensity scenarios. This

Table 5.2. Summary of Energy Information Administration (EIA) Projections for CO_2 Emissions and Gross Domestic Product (GDP), 2020

	Uncertainty Range between High and Low Projections, 2020 (*percentage points, relative to the reference case*)		Change in Projections, EIA 2001 v. EIA 1999 (*percentage point difference*)		
Country	**CO_2**	**Intensity**	**GDP**	**CO_2**	**Intensity**
United States	13.6	27.4	30.6	3.3	–20.9
Canada	21.1	20.0	5.7	–1.1	–6.4
United Kingdom	16.1	24.8	5.7	6.1	0.4
France	20.0	21.1	4.1	8.9	4.6
Germany	16.5	24.8	–7.3	–6.6	0.7
Japan	24.4	17.1	–9.9	–1.4	9.5
Former Soviet Union	42.0	48.4	35.5	14.9	–15.2
China	56.1	29.1	9.2	–17.1	–24.1
India	40.0	19.9	13.5	–3.8	–15.3
Korea (South)	36.6	26.9	2.0	–23.9	–25.4
Mexico	27.1	13.0	6.6	28.5	20.6
Brazil	45.8	14.7	8.1	13.4	4.9
Total Industrial	**16.7**	**24.4**	**9.0**	**3.5**	**–5.1**
Total Developing	**47.7**	**17.8**	**9.8**	**–5.4**	**–13.8**
Total World	**34.0**	**16.0**	**9.5**	**–0.6**	**–9.2**

Source: Compiled from EIA (2001b and 1999).
Notes: The uncertainty range is the percentage point gap between the high growth scenario and the low growth scenario compared to the reference scenario. EIA's change in GDP projections for the United States is not a typographical error. Long-term growth rates were increased from 2.2 to 3.0 percent per year.

"uncertainty range" is the percentage point gap between the high-growth and low-growth scenarios, compared with the reference scenario.

Future emission uncertainties (for the year 2020) are extreme in developing countries, with an uncertainty range of about 48 percentage points. This means that in 2020, according to EIA, emissions in developing countries could be anywhere between 3,490 and 5,697 million tons of carbon—a band of uncertainty that is larger than all developing country emissions in 1999 combined. When forecasts are expressed using an intensity indicator, this uncertainty is lowered to about 18 percentage points. In the case of Korea, the uncertainty (or range of forecasts) in absolute levels of CO_2 emissions is 36.6 percentage points, while the uncertainty range for the emissions intensity of CO_2 relative to GDP is 26.9 percentage points, a reduction of 9.3 percentage points. Figure 5.2 shows a comparison of the uncertainties, expressed in absolute levels of emissions and emissions intensities, for various developed and developing countries.

Future emission levels in industrialized countries, on the other hand, are less uncertain. Table 5.2 shows an uncertainty range of about 17 percentage points between the high and low CO_2 scenarios (relative to the reference case). It is interesting that future uncertainty in industrialized countries is actually greater when expressed using an intensity indicator than an indicator based on absolute emission levels (about 24 percentage points versus 17 percentage points, respectively).

Future projections for a given year (e.g., 2020) are also subject to continual, and sometimes major, revision over time. Lutter (2000) analyzes EIA's past projections of U.S. emissions and finds that nearly 87 percent of the forecasts turned out to be too low. Although the forecast errors are not that significant in the U.S. case, the situation for developing countries is different, as shown in Table 5.2. In the case of Korea, EIA forecasted GDP 2 percent higher in 2001 than in 1999, while the projection of CO_2 emissions decreased by 23.9 percent between the 2 years. As a result, the forecast of Korea's intensity decreased by more than a quarter. This phenomenon is not unique to EIA projections. In 1990, the Korea Energy Economics Institute (KEEI 1990) predicted that CO_2 emissions from the energy sector in 2010 would be 126.5 million tons of carbon. A decade later, KEEI (2000) predicted 2010 emissions to be about 170.6 million tons, an increase of about 35 percent.

Analysis of these forecasts—and comparisons between indicators—suggests several policy implications. First, CO_2 intensity targets (and dynamic targets in general) are likely to be superior to fixed CO_2 targets for developing countries, due to the reduced risk that the target would turn out to

Figure 5.2. Projected CO_2 Uncertainty in the Republic of Korea: Absolute Emissions versus Emissions per unit of GDP

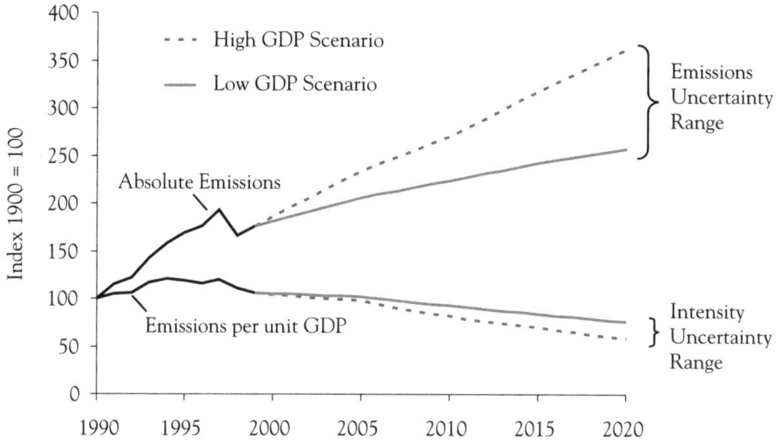

Source: World Resources Institute, compiled from data in EIA (2002a).
Abbreviations: GDP (gross domestic product).

be overly stringent. Second, fixed targets could be better for some industrialized countries or economies in transition. For industrialized countries, future projections of intensities tend to be even more uncertain than for emissions. Furthermore, analysis by Lutter (2000) shows that CO_2 emissions in larger economies are less variable from year to year. A 10-fold increase in the size of the economy leads to a decrease in variability of approximately 7 percentage points. On the basis of this, Lutter shows that forecast errors decline as the size of an economy grows.[2]

Finally, and more generally, uncertainties persist using both indicators and in all countries. There seems to be considerable uncertainty that is unavoidable, even with the intensity approach. Such uncertainties indicate that negotiators and advocates should be very cautious about using a BAU forecast as a benchmark or a baseline for determining targets, whether fixed or intensity.

In spite of the large uncertainties, future forecasts are indispensable in negotiating emission targets. The reduction burden of any given target is the gap between BAU and that target; this is related not to past performance but to the future commitment period. Therefore, the inherent uncertainty of BAU forecasts may harm the negotiations due to differing perceptions or estimations, not to mention possible strategic misrepresentation by Parties. Because dual targets accommodate a *range* of BAU forecasts, rather than a singular BAU point estimation, they might perhaps help negotiators reach agreement more quickly and easily. This topic will be revisited in Section III.

Regression Analysis: Case of the Republic of Korea

Using data from the Republic of Korea (South), this section illustrates the potential formulation of a dual-intensity target. Regression analysis is helpful because it assess how much certain factors (independent variables), in this case GDP, explain changes in emissions (the dependent variable).[3] Specifically, regression analysis is used here to determine the two intensity formulas (I_1 and I_2) as well as the GDP coefficient (α).

We derived a regression equation to relate emissions to GDP over the same time period. The results are summarized in Table 5.3. The R-squared value is 0.974, meaning that there is a strong relationship between the dependent (CO_2 emissions) and independent (GDP) variables. (An R-squared value of 1.00 would mean that 100 percent of the variation in emissions is explained by changes in GDP.) The coefficient for the GDP variable is estimated to be about 0.955, indicating an almost linear relationship between GDP and emissions. The equation derived produces a typical form of target emissions under the intensity approach as follows:

$$Emissions\ (t) = 1.239 \times GDP(t)^{\ 0.955}, \quad (1)$$

where *t* is the time frame for the commitment period, emissions(*t*) are the allowable emissions during the commitment period, and GDP(*t*) is the actual GDP during the commitment period. The above formula shows a GDP multiplier of 0.955, which is close to 1. The multiplier would be

Table 5.3. Regression Analysis of CO_2 Emissions in South Korea

Regression Formula:

$$Ln(Emissions) = Constant + \alpha \cdot Ln(GDP)$$

Adjusted R^2	Standard Error	constant	α
0.974	0.0664	0.2144 p-value (0.3170)	0.9554 (2.78×10^{-14})

Note: The regression analysis covers the years 1981 to 1998.

Box 5.1. Calculating the Values of the "Dual" Intensities

The standard error (difference between the realized emissions and the BAU forecast) from Table 5.3 is 6.6 percent. Assuming the forecast error has a normal probability distribution where the mean equals zero and the standard deviation equals the standard error, the 95 percent confidence interval for the model is then calculated as follows:

$$\text{projected emission} \times [1 \pm 1.96 \times (\text{standard error})]$$

An equation for the dual-intensity target can be derived from equation (1). Let us consider a situation in which a dual-intensity target for 2008 is predicted in 1998 on the basis of equation (1) and we want to limit the possibility of emissions turning out to be either higher than the compliance target or lower than the selling target by less than 5 percent. In other words, the possibilities of hot air and unintended reduction burden should not exceed 2.5 percent, respectively. By applying equation (1), we get the following result:

$$\textit{Emissions}\ (2008) = 1.239 \times \textit{GDP}\ (2008)^{0.955}, \quad (2)$$

An interval for the dual-intensity target for 2008 can be described as follows:

$$\textit{Intensity}\ (2008) = \frac{\textit{Emissions}\ (2008)}{\textit{GDP}\ (2008)^{0.955}} = 1.239 \times [1 \pm 1.96 \times 0.0664]$$

Intensity targets for compliance and selling can be derived from equation (2). The former equals 1.369 and the latter 1.109. In other words, Korea would be allowed to sell extra permits if the intensity defined in equation (2) turns out to be lower than 1.109 and is obliged to buy emission permits from abroad to ensure its intensity does not exceed 1.369. The number of permits Korea is allowed to sell in 2008 can be calculated as "actual GDP powered by 0.955 and multiplied by 1.109" minus "actual emissions." And the number of permits Korea is obliged to buy in 2008 is calculated by "actual emissions" minus "actual GDP powered by 0.955 and multiplied by 1.369."

different depending on different countries' economic situations. It is 0.5 in case of the Argentine Republic (1999, and Chapter 6 of this volume), and Lutter (2000) proposes a value of 0.6 for universal application to all countries. The multipliers could be developed and applied through in-depth, country-specific studies. Box 5.1 indicates how values for the dual-intensity targets are calculated. (Note that this box is not indicative of any future commitment by Korea and should be understood as a hypothetical example only. Furthermore, the year 2008 used in this example also should not be understood as any indication of the appropriate timing of commitment.)

Above, we derived dual-intensity formulas using *historical* data (1981–98) in a regression analysis. An alternative methodology might be to investigate the pattern between emissions and GDP under various *future* scenarios. According to EIA (2001b) projections, the rate of emission change tends to be larger than GDP in the low, reference, and high economic growth scenarios. This implies that the multiplier on GDP in the intensity formula is likely to be higher than 1 and that the GDP elasticity of emissions is greater than 1. However, this is not true in the case of the Argentine Republic (1999) and other countries. Therefore, additional in-depth analysis of the appropriate form for the intensity formula is needed.

III. Implementation Issues

As discussed in Section I, the main advantage of dynamic targets in general and dual-intensity targets in particular is that they can reduce economic uncertainty in the target-setting process, especially for developing countries where future uncertainties are more significant. However, the absence of a fixed environmental outcome under dynamic targets can create several implementation challenges. These challenges include interactions with international emissions trading, monitoring and verification of GDP, and complexity of the negotiating process. Except where noted, these issues are associated with dynamic targets in general and are not specific to dual-intensity targets.

Linkage to International Emissions Trading

To ensure environmental effectiveness and cost-effectiveness, a smooth interaction between emission targets and international emission trading is needed. Thus, the dual-intensity approach (and dynamic targets in general) needs to be fully compatible with emissions trading. It is also important for trades to be possible between countries (or private entities) using fixed targets and those using dynamic ones. Two key issues are associated with the compatibility of dynamic targets and international emissions trading.

The first issue concerns the overarching stability of the emissions trading system. Some of the "risks" of international emissions trading are systematically under-appreciated in climate change policy debates and may be challenging to manage in the future (Baumert et al. 2002). For instance, overselling of allowances, excessive uncertainty over market prices, and trading ineligibility could plague a trading system.

Given these risks, dynamic targets could offer an advantage over fixed targets, mainly in that dynamic targets are less prone to creating hot air, as discussed earlier. Lowering the risk of hot air (or, conversely, of overly stringent carbon constraints) could reduce the volatility of market prices by better balancing supply and demand, thus increasing liquidity. Dynamic targets also enhance market stability, in that some countries, under extreme circumstances, might be unwilling to comply with fixed targets that do not accommodate their economic realities.

The second compatibility issue between dynamic targets and trading is defining and managing the tradable unit. The tradable unit with dynamic targets is identical to that of fixed targets—"allowances" denominated in units of CO_2 (or carbon) equivalent. However, the quantity of these units available to a country is not known until the *end* of the compliance period because the allowable emission level is linked to actual GDP levels. This system differs from fixed targets, where the number of allowances is determined ahead of time and does not change. Thus, dynamic targets can add uncertainty and complexity to the emissions trading system.

There are several ways to enable trading to take place with dynamic targets. First, and most obvious, is a post-verification trading system, whereby transfers take place *after* emissions and GDP are verified (during a "true-up" period, such as the one adopted under the Kyoto Protocol). Here, some of the dynamic cost-reducing benefits of trading could be lost. However, earlier trades could take place through various derivatives (e.g., futures, options) and insurance contracts. A second way of addressing trading shortcomings is by determining the country's allowable emissions just prior to the commitment period, based on GDP projections for the commitment period. These projections could, in turn, be updated annually during the commitment period, and then reconciled at the end of the commitment period so that allowable emission levels reflect actual GDP changes.

Contrary to conventional wisdom, however, fixed targets encounter similar uncertainty: Countries do not know the total number of allowances available to sell (or needed to buy) ahead of time. Indeed, this uncertainty is structural. Because of time lags in determining actual emission levels, a country will not know its surplus or shortage of allowances until 2014 or later. Philibert and Pershing (2001) even state that, if the link between emissions and GDP holds, "the uncertainties on both will essentially compensate. In fact, the uncertainty regarding the availability of [allowances]...would likely be reduced, not increased, by dynamic targets in comparison to fixed targets."

Furthermore, the 2001 Marrakesh Accords contain numerous provisions suggesting that, with respect to emissions trading, targets are *already* dynamic. First, the Marrakesh Accords established a "commitment period reserve" system to guard against the risk of overselling. This commitment period reserve already envisions annual adjustments to a country's reserve level.[4] Second, the Accords created a "removal unit" (RMU) that can be issued annually on the basis of a net removal of GHGs that results from an approved set of activities (UNFCCC 2002, 59–60). RMUs will essentially increase a country's allowable emission level, yet the quantity of RMUs created will be known only *ex post* and will be subject to myriad requirements.[5] Provisions for RMUs and commitment-period reserves demonstrate that the amount of GHGs a country is allowed to emit and trade can already shift *during* the commitment period.

Given the wide-ranging uncertainties with respect to availability of tradable allowances, it is not clear that dynamic targets will pose additional problems, other than adding complexity. It is likely that a significant amount of trading activity will take place during the so-called "true-up" period during which emission quantities are known, RMU units have been issued, commitment-period reserves are solidified, and (in the event of dynamic targets) GDP values are established.

Finally, it should be noted that dynamic targets in general—and dual-intensity targets in particular—can also be used for S-CDM initiatives (see Chapter 3), should they be allowed under the Protocol. Under a sectoral dual-intensity approach, selling targets could serve as a baseline for generating tradable "certified emission reductions," which would make a compliance target unnecessary.

Gross Domestic Product: Choice of Currency

Along with emissions trading, the use of GDP as part of an emissions target has sparked some criticisms and concerns that should be taken seriously. The first relates to choosing a currency.

There are several ways to measure GDP. It is measured primarily in local currency; once this is done, it can then be converted into U.S. dollars (using market exchange rates) or international dollars (using purchasing power parities) to facilitate comparisons *across countries*. Also, each of these currencies (with the usual exception of PPP) can be measured in either "constant" or "current" terms (i.e., adjusted for inflation or not adjusted for inflation). The purpose of using constant currency is to facilitate comparisons *across time*.

Dynamic targets would, in all likelihood, express GDP in terms of *domestic* currency (Baumert et al. 1999). The first reason for this is that there is no need to compare intensities across countries. Comparisons of intensity levels across countries are not suggestive of the relative stringency of commitments, just as absolute emission levels were not used in Kyoto to gauge stringency. Rather, *percentage reductions* for each country, relative to historical levels or BAU, are typically compared to gauge relative stringency. The second reason is that values expressed in other currencies (such as U.S. and international dollars) are derivative of domestic currencies. Converting domestic currency would create unnecessary controversy regarding the proper exchange rate and PPP conversion factors. Also, the domestic currency would be expressed in *constant* terms because of the need to compare across time. The proposal by the Argentine Republic (1999, see Chapter 6, this volume) used constant domestic currency (1993 pesos) for calculating GDP.

Finally, what really matters for dynamic targets are annual rates of change, rather than absolute levels of GDP.[6] Rates of change are not strictly tied to currencies and might be easier to agree on and verify, since different measurement methodologies might yield the same rates of change. There is no need to engage in debates about what constitutes the "true" GDP of a country.

Gross Domestic Product: Monitoring and Verification

A second concern about GDP relates to monitoring and verification of GDP. Dynamic targets increase the data requirements for participating countries. Greenhouse gas emissions are already subject to a wide range of measurement standards, reporting requirements, and review provisions. If GDP were used to adjust emission targets, GDP would also need to be subject to scrutiny.

Generally, most countries and international institutions have more expertise on and experience with national economic statistics such as GDP than they do with measuring GHG emissions. The standards and methods for national income accounting have been developing for more than 50 years and are periodically updated by international institutions, such as the International Monetary Fund (IMF) and the United Nations Statistical Commission. The table in Appendix 5A offers a comparison of the systems for measuring, reporting, and verifying GHG emissions and GDP.

For each system that has been set up to account for GHGs, one or more analogous systems for GDP accounting are already in place. These systems need not be duplicated by the Climate Convention. In fact, the Conference of the Parties may, according to the Climate Convention, "seek and utilize, where appropriate, the services and cooperation of, and information provided by, competent international organizations and intergovernmental and non-governmental bodies."[7] The IMF, for example, could play a role in providing GDP data or verifying the data provided by countries through its existing "surveillance" and oversight processes.

Despite the availability of standards and oversight systems, many countries still do not report timely, internationally reliable GDP estimates (similar to the gaps in GHG reporting). The Milestone Assessments of the System of National Accounts show that many developing countries are not reporting GDP data. In addition, some countries, especially China, have been accused of purposefully inflating their GDP statistics. The mainstream press has repeatedly reported experts' suspicions that China overstates its economic growth (typically reported as 7 percent per year or more) to promote foreign direct investment.[8] In many countries, including China, statistical agencies are not functionally independent and can be subject to political influence.

Intentionally inflating, or "gaming," GDP is a legitimate concern because it would weaken emission targets. However, it is difficult to imagine that climate change policy could motivate such actions. GDP is used for a myriad other purposes, including by international organizations to determine eligibility for loans, aid, or other funds. GDP and derivatives of GDP (such as debt/GDP ratios) are used frequently as part of the terms and conditions for obtaining commercial loans. GDP also is used to determine financial contributions that support international institutions, such as the UNFCCC Secretariat. If a country wanted to cheat using a dynamic target, it would probably be more tempting to purposefully *understate* emissions rather than *overstate* GDP.

Overall, most emission reporting (under Article 12 of the Convention) is now insufficient to support binding emission targets; the same is true for GDP. If a country were to adopt a dynamic target, better reporting and independent verification (for which guidelines and institutions already exist) would be required for both emissions and GDP. This suggests the need to improve in-country capacity in both areas.

Inclusion of Non-CO_2 Gases and Non-Energy-Related Sectors

The analysis in Section II of this chapter includes only CO_2 emissions from fossil fuel consumption. For these emissions, correlations with GDP typically are extremely high. However, if a target included other gases and/or sectors (e.g., methane from agriculture), dynamic targets might not be as effective in reducing uncertainty. This is illustrated in Chapter 6, as Argentina's emissions from the agricultural sector typically did not adjust in response to GDP changes. Similarly, CO_2 from land use change (a major source in some developing countries) would likely correlate poorly with GDP.

This poor correlation suggests that precision in target setting will be even more elusive and uncertainty even harder to reduce. It also suggests a greater need for a dual-target approach to better account for these rampant uncertainties in the target-setting process.

Complexity and Capacity in the Target-Setting Negotiating Process

Generally, dynamic targets may make negotiations more complex, especially when attempting to differentiate commitments among many countries. Not only might counties adopt different percentage reduction commitments (as in Kyoto), they might also adopt different GDP adjustment provisions for targets (in other words, different α coefficients, in the case of intensity targets, or different emission adjustment percentages, in the case of indexed targets). Negotiations might become exceedingly complex, to the point that non-specialists, or indeed anyone other than the negotiators themselves, would have difficulty understanding proposed commitments.

With respect to *dual-intensity* targets, it is difficult to predict how the added complexity would affect the negotiating process. Using a *dual-target* concept might actually help countries reach agreement more easily. Negotiations would not need to reach a consensus on a *single* target; instead, they would focus on agreeing to the selling and compliance target intensities described above (I_1 and I_2). One can conceive of a two-step negotiating process under which a country proposes its own compliance target and the Protocol Parties collectively (or representatives from other countries) suggest the country's selling target.[9] Because the Convention requires consent from the country in question as well as Protocol Parties collectively, the distance between the two targets may converge to a reasonable level. Overall, dual-intensity targets make assumptions and political decisions more transparent during initial target-setting. This reduces the likelihood of surprises that might lead a country to defect from its commitment, possibly improving the prospects of agreement.

Complexity also points to capacity needs. Country delegations would need the training and skills to understand and assess various dynamic-target options. Thus, this approach might be best suited to the more advanced developing countries. To make things simpler, the negotiation process might benefit from an initial agreement on several different dynamic-target formulas to provide some standardization in methodologies (e.g., a few different GDP coefficients, α).

Determining the Stringency of Reduction Commitments

Some approaches to target setting, such as the Brazilian Proposal and per capita entitlements, include provisions for determining the proportionality of emission limitation requirements among countries (Chapters 7 and 8). (For the examples noted above, these provisions are based on relative responsibility for existing climate change and on population size, respectively.) In other words, the stringency of a country's reduction commitment is partially[10] determined by the approach itself. Dynamic targets are different in that the stringency of the reduction target is separate from the approach. Generally, the stringency of such a target is an equity issue. This topic begins to exceed this chapter's scope, not because it is less important, but because it is more political than theoretical and could be considered independently without altering the essential elements of this approach. In principle, a variety of equity criteria could be applied to an intensity-target approach in order to determine the stringency of country targets. This approach, however, would most likely be employed through a pledge-based process, whereby countries suggest their own target(s), and negotiate this target(s) with the rest of the Parties.

Nevertheless, several proposals for dynamic targets do address the issue of how to determine the stringency of short-term reduction targets. CCAP (1998) suggested a growth baseline where the target intensity is set to be lower than the BAU level but higher than an intensity that can be achieved through "no-regrets" measures. They also suggested that countries of similar circumstances could be grouped together, with a common rate of intensity improvement required of all countries in a particular group. Four criteria would be considered in defining the groups: fuel mix, economic growth, technology level, and policy framework.[11] Frankel (1999) points out that a dynamic target set at BAU levels would have environmental and economic benefits for all countries involved (assuming, of course, the existence of an international market for emission reductions). He suggests that negotiations could settle near a "break-even" level, where overall gains from trade equal overall domestic costs, and that richer developing countries could take deeper targets than poorer ones.

In the case of Argentina (see Chapter 6), the target implied an emission reduction between 2 and 10 percent across the assessed scenarios. Besides Argentina, the United States is the only country to propose a dynamic target, albeit a non-binding one that is not linked to international emissions trading. The U.S. proposal of an 18 percent voluntary reduction in GHG intensity, announced in 2002, suggests future emission levels that are similar to historical trends (WRI 2002), implying little, if any, additional effort.

Internationally, it is unlikely that a single rule could guide the target-setting process. To make it fair, polluter-pays and egalitarian principles could play a role in burden sharing, and therefore cumulative and per-capita emissions could be used to help determine target intensities. Ability to pay and capacity to reduce could also play a role in target setting; thus, per capita GDP and marginal abatement cost characteristics would be considered. Marginal abatement costs are widely known to vary significantly among countries.

IV. Conclusion

The usefulness of dynamic targets and dual targets depends on the problem to be solved. In the past, countries have been extremely concerned with the magnitude and attendant economic impacts of taking an emission target. Yet, determining the economic impact of a fixed target is hard. Targets are negotiated 10 to 15 years in advance of their implementation, making it extremely difficult to gauge the level of effort inherent in any single target. Negotiating emission controls is challenging precisely because of pervasive uncertainties: Countries do not actually know what they are agreeing to. Dynamic targets and dual targets (perhaps combined) may have compelling advantages over fixed targets in that they can help reduce the problem of *uncertainty*.

Similarly, in negotiating future emission targets, a developing country might want to be protected from the possibility of having to be a net buyer of emission reductions.[12] Here, dual targets could be especially useful and could be combined with fixed or dynamic targets. The compliance target, on the one hand, could be set conservatively (or not at all) to ensure that under most any scenario, the country's BAU emissions would not exceed that level. The selling target, on the other hand, could be set more stringently to create an incentive to reduce domestic emissions (and capture benefits through international emissions trading).

An inevitable consequence of using dynamic or dual targets is a lack of environmental certainty. Considering the long-term nature of the climate change issue, however, short-term environmental certainty may be less important than the overall stringency of the reduction target. If an intensity target could provide more stringent reduction objectives, it may be more desirable to have such a stringent target even with the attendant lack of environmental certainty.

Key conclusions of this chapter include the following:

- Dynamic targets can be more effective than a fixed target in reducing the risk of "hot air" from weak targets or of non-compliance from unintentionally burdensome targets. This conclusion holds primarily for many developing countries, although not necessarily for industrialized countries. In industrialized countries, projecting *absolute* emission levels might be more reliable than projecting intensities.
- Dual targets could further reduce the risk of undesirable hot air or non-compliance. The concepts of dual and dynamic targets can be combined through the use of dual-intensity targets, for example. Dual targets could also be used with fixed targets.
- Dual-intensity targets would increase the complexity of the negotiations. Paradoxically, however, they might also facilitate a ratifiable consensus. Again, this conclusion applies primarily to negotiating developing country targets and not necessarily industrialized countries.
- Dual-intensity targets address only one part of a climate protection architecture. Other elements also are integral to the overall framework. They include monitoring and verification of both GDP and emissions data. International emissions trading which, we believe, is sufficiently compatible with dynamic targets, is also critical to an international policy framework.
- Dynamic targets are not a burden-sharing approach. Rather, they are a way of shaping a target. To promote the real application of this approach, additional decisions would need to be made on the acceptable stringency of country targets.
- As with almost any approach, to make dynamic targets operational, serious country-level analysis is required (including decisions on gases and sectors to be covered), as Chapter 6 of this volume illustrates.

For many developing countries whose unstable and uncertain economic growth exacerbates emission uncertainties, dual-intensity targets may be the best option—a low-risk strategy for participating fully in global climate protection.

Appendix 5A.

Table 5A. Measurement, Reporting, and Review of Information: Greenhouse Gases and Gross Domestic Product

	Greenhouse Gases	Gross Domestic Product
METHODOLOGIES AND STANDARDS	**Kyoto Protocol (Art. 5, par. 2)**: Requires the use of emissions (and absorption) estimation methodologies that are accepted by the Intergovernmental Panel on Climate Change and agreed on by the Conference of the Parties. **Intergovernmental Panel on Climate Change (IPCC)** provides guidelines and good practice methodologies for estimating greenhouse gas emissions.	**System of National Accounts (SNA)** • SNA is a common set of concepts, definitions, classifications, and accounting rules used in economic analysis and policymaking for all countries. The SNA provides a comprehensive conceptual and accounting framework for analyzing and evaluating economic performance. • Updated periodically through a working group that, to ensure consistency and comparability, includes the United Nations, Statistical Commission, the International Monetary Fund (IMF), the Organization for Economic Cooperation and Development (OECD), Eurostat, and the World Bank. **IMF Article IV consultations.** Data gathering through Article IV consultations relies on an internal IMF process and it responds to specific informational needs of the IMF such as for data on gross domestic product (GDP). **IMF's Special Data Dissemination Standards (SDDS)** guides countries in the dissemination of financial statistics (in order to promote access to international capital markets). The SDDS includes standards in the following areas (1) data: coverage, periodicity, and timeliness (or reporting); (2) public access to data; (3) integrity of the disseminated data; and (4) quality of the disseminated data.
REPORTING	**United Nations Framework Convention on Climate Change (Art. 12).** Periodic reporting of national communications, including a national emissions inventory **Kyoto Protocol (Art. 7).** Annual emissions inventories and necessary supplementary information to ensure compliance.	**SNA.** The U.N. Statistical Commission sends an international questionnaire to be filled out by members voluntarily on an annual basis. **IMF Article IV Consultations** (surveillance), contrary to the 1993 SNA, a member country (of the IMF) has the obligation to provide the information requested by the IMF's staff as stated in IMF's Article IV. The country itself, though, decides the public availability of this information to avoid the disclosure of sensitive information. IMF surveillance activities are conducted annually. **SDDS**. See above.

continued next page

Table 5A. *continued*

	Greenhouse Gases	Gross Domestic Product
MONITORING, REVIEW AND VERIFICATION	**Kyoto Protocol (Art. 8).** The information submitted by each Annex I Party shall be reviewed by expert review teams	**IMF Article IV Consultations.** See above. IMF **Reports on the Observance of Standards and Codes** assess the extent to which countries subscribing to the SDDS observe international standards **Milestone Assessment of the Implementation of the SNA** is a system for monitoring and assessing the performance of countries. The system includes six milestones that indicate different levels of national accounts development. Generally, the **SDDS** promotes dissemination, transparency, and public access to data. These data can then be reviewed and assessed by financial institutions (e.g., creditors) and others.
CAPACITY BUILDING	**National Communications Support Programme.** Provides technical support to enhance the capacity of non-Annex I parties in preparing their initial national communications, including in the preparation of greenhouse gas inventories. **CC:Train.** Jointly created by the Climate Convention Secretariat and the United Nations Institute for Training and Research in 1994.	The IMF's **General Data Dissemination Standard** (GDDS) focuses on education and training to improve data quality. GDDS includes a process for needs evaluation for data improvement and priority setting. Nine regional seminars for country officials have been held to date. IMF **Reports on the Observance of Standards and Codes** aim to assist countries in identifying areas where transparency can be further enhanced. **IMF Article IV Consultations.** See above.

Notes

1. The case of Russia under the Kyoto Protocol illustrates both kinds of hot air—intentional and inadvertent. At the time of negotiation, it was probably envisioned that Russia's economy would recover more rapidly.
2. Lutter (2000) argues that for forecast errors that are one period ahead, a 1 percent increase in GDP is associated with a reduction in the forecast error of about 0.1 percent.
3. This analysis uses "reduced form" regression models of the log-linear equation, using data from the IEA (2000) on two key variables, CO_2 emissions and GDP, for the period 1971 to 1998.
4. UNFCCC (2002, 54): "Each Party included in Annex I shall maintain, in its national registry, a commitment period reserve which should not drop below 90 per cent of the Party's assigned amount calculated pursuant to Article 3, paragraphs 7 and 8, of the Kyoto Protocol, or 100 per cent of five times its most recently reviewed inventory, whichever is lowest."

5. UNFCCC (2002, 62–63). Regarding restrictions on RMU eligibility, consider the uncertainty inherent in the following chain of conditional requirements: "Each Party included in Annex I shall issue in its national registry RMUs equivalent to the net removals of anthropogenic greenhouse gases resulting from its activities under Article 3, paragraph 3, and its elected activities under Article 3, paragraph 4, accounted in accordance with decision -/CMP.1 (Land use, land use change and forestry) as reported under Article 7, paragraph 1, following completion of the review in accordance with Article 8, taking into account any adjustments applied in accordance with Article 5, paragraph 2, and resolution of any questions of implementation related to the reported net removals of anthropogenic greenhouse gases. Each Party shall elect for each activity, prior to the start of the commitment period, to issue such RMUs annually or for the entire commitment period."
6. Indexed targets explicitly use rates of change, and intensity target formulas could be algebraically rewritten to use rates of change.
7. Article 7, paragraph 2(l).
8. For example, see Waldron (2002).
9. From the game-theoretic perspective, one can see a strong incentive for a country to increase both selling and compliance targets. A game rule needs to be designed to mitigate such strategic behavior. The proposed rule is one example where the bargaining power is distributed such that the country in question is given a primary opportunity to set its compliance target (which may determine the financial burden in case of lower performance) and other countries (e.g., the COP/MOP) have the role of setting the selling target.
10. Of course, the stringency of the targets also is partially determined by the overall environmental goal.
11. CCAP (1998) classifies 12 high-emitting developing countries into five categories. China and Iran are included in the group with "high no-regrets potential" and South Africa is classified in the "medium-high" group. India, Indonesia, Saudi Arabia, South Korea, Taiwan, and Thailand are classified in the "medium" potential group, and Mexico and Venezuela are evaluated as having "medium-low" potential. Brazil is classified in the "low potential" group.
12. This scenario is, of course, only viable if industrialized countries are likely to be net sellers.

6. Learning from the Argentine Voluntary Commitment

Daniel Bouille and Osvaldo Girardin

Introduction

An unprecedented event took place at the Fourth Conference of the Parties (COP 4) to the United Nations Framework Convention on Climate Change (UNFCCC) in 1998 in Buenos Aires. The host-country president, Carlos Menem of Argentina,[1] announced his government's commitment to establish a voluntary greenhouse gas (GHG) emissions target for 2008 to 2012 and to formally commit to this target the following year during COP 5. This was the first time a developing (non-Annex I) country had agreed to meet a quantified GHG limitation target.

However, the two principal legal instruments of the climate change regime—the Climate Convention and the Kyoto Protocol—have not established provisions for voluntary emissions targets. This poses numerous questions for the future of climate negotiations. How should voluntary commitments be incorporated into the UNFCCC and Kyoto Protocol? What are the advantages and disadvantages of voluntary commitments for countries not immediately obligated to take targets, particularly in light of reticence on the part of industrialized countries to implement initiatives to meet their own commitments? What are the economic and environmental implications of the Kyoto Protocol mechanisms—specifically, the Clean Development Mechanism (CDM)—for a country making a voluntary commitment, considering the uncertainties about the size and impact of the carbon market for a developing economy?

This chapter synthesizes and analyzes the process through which Argentina developed its voluntary commitment, focusing on the technical aspects of the proposal, the likelihood of effective implementation, the target definition process, and the level of participation of different actors in this process.

Section I of this chapter examines the national and international context in which the commitment process took place. Section II explains the different target types (i.e., fixed and dynamic) and emission reduction levels considered by Argentina. This section also explores whether the target methodology would be useful to other developing countries. Section III focuses on the main implications of adopting this target within the framework of the Climate Convention and Kyoto Protocol, as well as the relationship to development priorities in Argentina. Section IV presents conclusions and lessons learned.

I. The Decision: Process and Justification

To understand why Argentina proposed a voluntary commitment, it is necessary to examine both the international and domestic pressures facing the COP 4 host country.

International Context

Since the Climate Convention's adoption in 1992 industrialized (Annex I) countries have pressured developing (non-Annex I) countries to make quantified emissions commitments.[2] In July 1997, the U.S. Senate adopted the Byrd-Hagel Resolution, which placed two conditions on any U.S. ratification of a binding protocol: The agreement must not threaten the U.S. economy, and "key developing countries" had to take on binding targets to limit emissions during the same commitment period.[3] The Kyoto Protocol's adoption in December 1997 further increased the pressure on developing countries to agree to emission limitation targets.[4]

Why did Argentina propose a voluntary target to limit GHG emissions? Argentina's foreign policy during the Menem Administration aimed at developing a closer alliance with the United States, which was evidenced by deepening bilateral relations and support of U.S. international policy. Argentina's adoption of a voluntary target to limit GHG emissions must be understood within this context, in which pressures on developing countries to voluntarily commit were combined with Argentina's foreign policy goals. Within the context of bilateral negotiations, this entailed a significant alliance with U.S. foreign policy in diverse forums and, regarding climate change, the Argentine proposal appears to be designed to explicitly support the U.S. position.

President Clinton's official visit to Argentina in October 1997 represented a major milestone in cementing this relationship. Presidents Clinton and Menem signed the Presidential Declaration of Bariloche, which proposed cooperation between the United States and Argentina on environmental matters, including global climate change.[5] This declaration underscored the need to establish a realistic and obligatory target based on flexible cost-effectiveness criteria, as well as the need to mobilize private sector resources for economic development projects in developing countries. These "joint implementation" (JI) projects would permit the reduction of GHG emissions. Both parties recognized that the response to climate change must be global and that all countries (developed and developing) should be involved.[6]

During the Kyoto negotiations (COP 3), Argentina's representatives agreed to promote voluntary commitments for developing countries in the Protocol.[7] However, there was no consensus on the so-called Article 10, which called for such voluntary commitments. Specifically, the Group of 77 (G-77) and China—the main developing-country negotiating bloc—found this provision categorically unacceptable. The proposal to include voluntary commitments was revisited the next year during COP 4,[8] although in the end it was not included in the conference agenda or discussed during the meetings leading up to the conference (UNFCCC 1998). Again, this opposition was led by the G-77 and China.[9, 10]

Argentine delegation officials later stated that they did not expect that their proposal on voluntary commitments would be accepted. However, they wanted to ensure that it was discussed as a way to free up negotiations and overcome the arguments against ratification of the Kyoto Protocol presented by Annex I countries.[11] From the start, the proposal's political and legal feasibility was dubious, since it required an amendment to the Protocol, which was not widely supported and not legally possible given that the Protocol had not yet entered into force.

Domestic Context

Domestically, the only actors in Argentina who entirely supported the adoption of an emissions target were those represented by political and technical officials who both made the decision at the highest levels and designed the proposal, as well as some civil society individuals. These actors, within the then-Secretariat of Natural Resources and Sustainable Development (SRNDS), strongly backed the announcement, development, and adoption of the target. The main argument they presented was that the voluntary commitment would open up the possibility of gaining access to all the flexibility mechanisms of the Kyoto Protocol. This misunderstanding about the accessibility of all the mechanisms to a developing country may have been the main error in the call for a voluntary target.

In internal discussions, national authorities argued that Argentine competitiveness in one of the Protocol's flexibility mechanisms, the CDM, was limited vis-à-vis the large non-Annex I emitters, such as India and China. However, two other mechanisms, emissions trading and JI, could provide access to markets in the medium to long term. By adopting a voluntary commitment, Argentina sought to gain access to emissions trading and JI—which were designed exclusively for Annex I countries—to attract investment, create new employment, and improve local environmental conditions without losing its non-Annex I status.[12] The magnitude of such investment was, however, unknown. This proposal became known as the "third way." Some Annex I countries, such as Canada, expressed interest in it, although recognizing that the rules for this third way had yet to be negotiated (Government of Canada 1999).

It is unlikely that the Argentine proposal stemmed from sustainable-development interests. Rather, Argentina was mainly interested (however erroneously) in accessing all of the mechanisms to overcome what was perceived as an obstacle to developing countries in the Protocol, and thus gain access to the derived business opportunities. In addition to the private sector, some NGOs were interested in adoption of the emissions target. The NGOs believed that the commitment of Argentina, an important country in the region, could be helpful to other developing countries in evaluating voluntary target taking.

The main national actors that were consulted for this evaluation of Argentina's decision[13] viewed the adoption of the commitment as an isolated measure. They believed the target was not part of an integrated plan linking climate change to other local, regional, and global environmental issues, nor was it part of a strategy for general and sectoral policies. Rather, most of them agreed that the Menem Administration's position was motivated by the aforementioned desire to establish a close alliance with the United States in terms of international policy. If there was a plan, it was based on vague principles and proposals regarding the advantages of avoiding conflict with the United States and developing actions that showed that Argentina was a strategic ally in international policy.[14]

At the time, academics and some industry sector representatives expressed their reservations about the establishment of a voluntary emissions target for the first commitment period. They did not see the point of such action when most of the main Annex I countries still had not made any significant advances in compliance with the Kyoto Protocol and had not even ratified it. They also did not agree with the decision to link the target to an aggregate indicator such as gross domestic product (GDP), given the large share of Argentine emissions coming from the livestock sector. The growth of this sector depends more on international prices and exchange rates than on GDP growth. Also, the supply structure of the Argentine energy sector had a low carbon intensity.

Establishing the Target

The process of quantification and adoption of the emissions target had two phases. First, the decision for Argentina to adopt a target was made at the highest levels of government without consulting other interest groups or even other areas of the government. It was made behind closed doors and within the President's closest circle. Second, with the promulgation of Decree Number 377/99 on April 16, 1999, work on the target type and quantification began through creation of the National Commission for the Elaboration and Proposal of a Greenhouse Gas Emissions Target. The Commission fell under the purview of the SRNDS of the Presidency.

The SRNDS, headed by Maria Julia Alsogaray, was the clear leader in this process, as it had the highest political backing and represented the position closest to that of the Clinton Administration. Participation by other government entities varied significantly. Those most engaged were the Secretary of Foreign Relations and Latin American Affairs (in the Ministry of Foreign Relations, International Trade, and Culture); the Energy Secretary and the Secretary of Agriculture, Livestock, Fisheries, and Nutrition (both within the Ministry of the Economy and Public Works and Services); and the Secretary of Science and Technology (in the Ministry of Culture and Education). The other government entities involved[15] did not participate in any significant manner, as was the case for representatives of provincial governments, indicating the lack of climate change and environment-related issues on the agendas of diverse public entities at all levels—national, provincial, and municipal.

The private sector participated through an Advisory Committee, made up of representatives from the business sector, science, and academia (both public and private), as well as from NGOs specializing in issues linked to the Commission's goals.

A technical team was formed to carry out baseline studies to elaborate targets for various sectors, principally the calculation of the GHG inventory for 1997, the elaboration of socioeconomic scenarios and sectoral emissions, and the identification of mitigation options. The technical team presented the partial and final results of these studies to the Commission and the Advisory Committee in different meetings. Their comments and suggestions were incorporated into the studies in some cases.

II. Taxonomy of the Target

Argentina's target aimed to maintain net anthropogenic GHG emissions (measured in tons of CO_2 equivalent) at a level that did not exceed the quantity defined as the "National Emissions Target." This target was expressed by the equation $Ep(t) = K * \sqrt{GDP(t)}$, where the emissions target is equal to a constant (K = 151.5) multiplied by the square root of GDP (at 1993 prices), averaged over the 5-year commitment period. If implemented, the target would deliver an emission reduction between 2 and 10 percent (relative to business as usual, BAU) depending on actual GDP growth and other factors affecting the base scenario.

Between March and August 1999, the technical team advising the Commission and Advisory Committee carried out studies to support the definition (i.e., the type), quantification (i.e., the magnitude), and decision to reach the mitigation target in compliance with Decree 377/99. They began formulating the target with an elaboration of the business-as-usual, or base, scenarios. First, based on information on GHG emissions in 1990 and 1994, they developed a simplified inventory for 1997. Second, they developed future socioeconomic scenarios (i.e., GDP projections). The agricultural sector (including livestock) required special consideration. In 1997, it accounted for more than 40 percent of Argentina's total emissions, and its growth is not dependent on GDP changes but rather more closely tracks exchange rates and international prices (Argentine Republic 1999).[16] Thus, activity-level projections were developed separately for the agricultural sector for the 2008–12 period. This exercise produced nine scenarios reflecting the entire spectrum of different emission possibilities given different GDP and agricultural sector assumptions (see Table 6.1).

To identify ways of reducing emissions, the technical team carried out mitigation studies in the energy, agriculture, forestry management, and waste sectors. These studies produced absolute and relative estimates of potential reduction or increase in sequestration of GHGs, all essential information to define the voluntary commitment.

The Commission chose the type and level of target on the basis of information from inventories, emission projections, product research (sectoral and aggregate), and mitigation-options studies considering both magnitude and feasibility.

In deciding which approach to use and in determining the level of the target, the Commission considered the following key factors: (1) the behavior of the agricultural sector (for the reasons noted above), (2) the need to avoid creating an obstacle to sustainable development, and (3)

Table 6.1. Estimated Greenhouse Gas Emission Reductions for Argentina Under Three Different Targets, 2008–2012

Growth Scenarios for GDP and Agriculture	BAU Emissions, annual mean value MtC	*Change in Emissions Relative to Baseline Scenarios* Fixed Target MtC (percentage)	Intensity Target MtC (percentage)	Square Root Target MtC (percentage)
Low GDP–Low Agro	95.6	–4.82 (–5.0)	8.03 (8.4)	1.89 (2.0)
Medium GDP–Medium Agro	105.2	4.81 (4.6)	4.64 (4.4)	4.85 (4.6)
High GDP–High Agro	122.3	21.87 (17.9)	–0.24 (–0.2)	11.54 (9.4)
Low GDP–Medium Agro	96.0	–4.38 (–4.6)	8.48 (8.8)	2.33 (2.4)
Low GDP–High Agro	102.4	1.96 (1.9)	14.82 (14.5)	8.67 (8.5)
Medium GDP–Low Agro	104.8	4.37 (4.2)	4.19 (4.0)	4.41 (4.2)
Medium GDP – High Agro	111.6	11.16 (10.0)	10.98 (9.8)	11.20 (10.0)
High GDP–Low Agro	115.5	15.09 (13.1)	–7.03 (–6.1)	4.76 (4.1)
High GDP–Medium Agro	115.9	15.53 (13.4)	–6.58 (–5.7)	5.20 (4.5)

Source: Derived from Argentine Republic (1999).
Notes: The "reference case" scenario is in bold. **Abbreviations:** BAU (business as usual), MtC (millions of tons of carbon equivalent).

the need to avoid generating "hot air."[17] Table 6.1 shows the nine scenarios used in the Argentine target analysis, as well as the three different types of emissions targets considered. Each scenario represents a particular set of assumptions about growth in GDP and the agricultural sector.

In any of the above scenarios, emissions are expressed by the equation[18]

$$E_{BAU\,(t)} = \alpha * GDP(t) + \beta * A(t) \quad (1)$$

where $E_{BAU\,(t)}$ is the emission level resulting from a lack of mitigation strategies (i.e., BAU), A(t) is the indicator of agricultural activity (prices and access to international markets), and α and β are indicators of the intensity of emissions.

By multiplying the emission intensities (α and β) by activity levels (GDP(t) and A(t)), the equation yields a business-as-usual emission level, expressed in tons of CO_2 equivalent.

The reduction of emissions can be expressed as

$$RE(t) = E_{BAU\,(t)} - E_{p(t)} \quad (2)$$

where RE(t) is the reduction of emissions and $E_{p(t)}$ is the emission level permitted to meet the target.

$E_{p(t)}$ can be defined in various ways. In Argentina's case, the Commission initially evaluated two possible kinds of target:

1. A fixed target expressed as an absolute reduction.
2. A dynamic target based on an index of emission intensity.

The fixed-target option required a "reference scenario" to effectively monitor compliance with the commitment. In the Argentine case, this reference scenario is median GDP growth and high activity in the agricultural sector. The fixed-target approach implies defining the volume of permissible emissions ($E_{p(t)}$) and setting it as a ceiling for all scenarios.[19] It was estimated that reductions of up to 10 percent could be achieved in this scenario, where total GHG emissions were equivalent to 111.6 million tons of carbon equivalent (MtC). The third column of Table 6.1 shows the results of adopting a fixed target under each of the nine scenarios considered. However, the adoption of a fixed target could entail two difficulties. Low-growth scenarios might generate phantom reductions ("hot air"), while high-growth scenarios would require very sharp emission reductions (in percentage terms).

Given these disadvantages, the Commission considered another option: a dynamic target, expressed in emission intensity (i.e., the ratio of emissions to GDP). Internationally, various emission intensity proposals have been developed, including one by the Bush Administration in February 2002. A target based on emission intensity is attractive because it does not entail limits on growth and also assumes "growing efficiency" in the socioeconomic system, thus generating additional benefits. Therefore, given a gradual percentage reduction of the intensity coefficient, the absolute magnitude of allowable emissions (and hence reductions) then depends on the level of activity.

In the Argentine case, the problem with an emission intensity target is related to the emissions from the agricultural sector which, as noted, tend to change in response to factors other than GDP. A target linked solely to GDP could introduce a serious risk of non-compliance. If GDP turned out to be relatively low, while agricultural activity turned out to be high, trying to comply with the target would exert significant socioeconomic pressures. At the same time, important changes in the GDP's structure could have devastating effects on the emissions commitments or on achieving greater efficiency. In this case, the risk of generating "hot air" is low, while greater GDP growth would facilitate compliance with the commitment. On the other hand, if the intensity of emissions is established as a fixed relationship with respect to the reference scenario, the result may be permissible emissions that generate hot air (if GDP growth is high, and agriculture activity is low). The fourth column of Table 6.1 shows the results produced by this alternative.

In developing countries, variations in the GDP's sectoral structure make it difficult to adopt dynamic commitments based on an aggregate economic measure such as GDP. Other difficulties involve the difference in the intensity of emissions of these sectors, as well as the links to and dependency of certain sectors on international markets.

Given the difficulties posed by Argentina's economic and emissions structure, technical experts explored other options for a dynamic target and developed a proposal based on an emissions index and square root of GDP. According to Barros and Conte Grand (1999), "The adoption of an index that utilizes the square root of the GDP means that the index itself empirically adjusts to the evolution of past Argentine emissions, as well as projected future emissions." In other words, the Commission selected the index to meet the objective of defining a formula that both accurately accounted for past emissions and produced reductions of up to 10 percent of future emissions.

The equation was formulated as follows:

$$E_{p(t)} = K * \sqrt{GDP(t)} \quad (3)$$

where

$$K = (1-\rho)\, E^{R}{}_{BAU\,(t)} / \sqrt{GDP^{R}(t)},$$

ρ is equal to the percentage of emission reductions with respect to the reference scenario E^R, and $E^{R}{}_{BAU\,(t)}$ is the emission level in the reference scenario.

This equation can be applied to any of the nine scenarios of growth in GDP and agriculture, but probably is only relevant to Argentina. In the case of Argentina, $\rho = 0.1$ and the resulting emission reductions varied between 2 and 10 percent, while total emissions grew between 22 and 44 percent, relative to 1997 levels. The intensity of emissions would be reduced between 12 and 25 percent compared to 1997, depending on the socioeconomic scenario (Barros and Conte Grand 1999).

The Argentine formula is an option tailored to match local circumstances, particularly regarding the agricultural sector. It could not be applied in other countries except as a model for adapting emission intensity to a specific situation based on relevant criteria.

The reference scenario (medium GDP growth and high agricultural sector growth) results in a 10 percent reduction in emissions compared to the BAU emission level. This target would allow Argentina to comply with the commitment (based on identified mitigation options), generate tradable gains, and avoid the creation of hot air.

The equation, its formulation, and its K value were chosen not on principle but on output, as some authors and expert observers have argued. It is difficult not to subscribe to this point of view given that the technical experts were tasked with finding a formula that accomplished the following:

- Reflected the historical evolution of emissions and their relation to GDP.
- Achieved a 10 percent (approximate) reduction with respect to the most probable or desirable scenario (shown in bold in Table 6.1).
- Guaranteed that under no scenario would hot air be generated.
- Guaranteed that under no scenario would the percentage of emission reductions be greater than 10 percent.

The formula taking the square root of GDP as the variable is the one that meets all these conditions, as long as the constant parameter (K) is 151.2.

The volume of allowable emissions for the reference scenario (10 percent (ρ) reduction in relation to BAU) was predetermined. Given the results of economic and agricultural scenarios, K and the relation with Ep(t) that guarantees a 10 percent reduction from the reference case while generating positive values that do not exceed 10 percent for the rest of the scenarios, were thus established. The incorporation of additional scenarios based on other hypotheses could lead to different values for K and a different relation to GDP. Therefore, the equation could be adjusted and used for a series of specific hypotheses and not just for the Argentine case, as can be inferred from the spectrum of scenarios and the combination of possible options for GDP's evolution and agricultural-sector activity levels.

III. Implications of the Adoption of Emissions Targets

Linking the Target to the UNFCCC and the Kyoto Protocol

As discussed, neither the Kyoto Protocol nor the UNFCCC contains provisions for voluntary commitments, such as the one presented by Argentina at COP 5. Argentina's announcement did not include a strategy for legally operationalizing its target within the international negotiation process. Yet, the voluntary commitment itself was conditional on an action by the COP "to present a new alternative that empowers non-Annex I countries, which like Argentina would like to adopt an emissions target, to participate in the mechanisms established in Articles 4, 6, and 17 of the Kyoto Protocol" (Barros and Conte Grand 1999). Two possible options for making the commitment operational included an amendment to the Kyoto Protocol or the consideration of other legal instruments (i.e., another Protocol) possibly involving another negotiating phase. Neither option was promising.[20]

The only way for Argentina to take a target that would grant it access to the mechanisms was by joining Annex I of the UNFCCC and Annex B of the Kyoto Protocol. In addition to access to JI, emissions trading, and the "bubble" (Article 4 of the Protocol) mechanisms, this could entail some advantages, such as additional "flexibility" similar to that accorded to economies in transition. Yet the drawbacks were significant, and would entail a long delay. Procedurally, entry into Annex I would require amending the Convention (Article 16) or declaring Annex I status under Article 4.2(g). The Protocol would also have to be amended for Argentina to join Annex B. Article 21 of the Kyoto Protocol states that amendments to Annex B (such as accession) must be adopted at a meeting of the Protocol Parties, which cannot occur until after the Kyoto Protocol enters into force. This requirement would entail a considerable delay for the effective application of the amendment and undermines the argument that Argentina's commitment was presented as a way to contribute to the ratification of the Kyoto Protocol.

Overall, the benefit of accession to Annexes I and B is not evident. In addition to the above difficulties, inclusion in Annex I also implies exclusion from the CDM, which is the only mechanism that allows credit accumulation before 2008.[21] Moreover, gaining Annex I status may lead others to question Argentina's receipt of foreign aid as a developing country. Proposing an amendment would also be politically problematic, since other developing countries opposed voluntary commitments. Many analysts believed that international negotiating conditions for developing countries would worsen if discussions were to begin on possible amendments to the Protocol and Convention.

Fortunately, the officials who formulated this target did not contemplate this option. The difficulties and uncertainties associated with existing options motivated Argentine officials to explore an entirely new approach, which they called the "third way." Voluntary target taking was seen as a potentially more attractive route for developing countries to enjoy some of the advantages of the Kyoto Protocol, without losing their status as non-Annex I countries.

The officials who set Argentina's target believed it was possible to lead a group of countries interested in establishing emission limits over a long period of time. However, these officials did not take into account that this would significantly change the Kyoto Protocol architecture, and that their idea could not be implemented without a radical modification of the treaty or an agreement on a new protocol. It implied a "flexibility of the flexibility mechanisms" to allow these countries to have access to all three mechanisms, not just the CDM. The hope that a subgroup of countries within the Kyoto Protocol and the UNFCCC would take on voluntary commitments (and perhaps establish another Annex) never materialized. Even if other countries were interested, there was still no procedure for altering the architecture of the Protocol prior to its entering into force.

The idea was not properly developed or explored in depth. Since there are no prior experiences in the UNFCCC or the Kyoto Protocol with this type of approach, it would require considerable effort to develop and implement within the framework of international negotiations. In the end, the voluntary target was shelved. The COP has neither accepted nor rejected it.

Long-Term Development Priorities

It is not very easy to identify the consistency between Argentina's development priorities and the setting of the emissions target. There are at least three reasons for this inconsistency.

First, the decision to commit to limiting emissions was based on international political objectives, without an analysis of the national economic impact. Second, the government's development priorities and policies have never been explicitly and systematically defined in strategic plans and planning processes. This has meant that climate protection and sustainable development lack any identifiable policy linkages. Third, within the context of a prolonged economic crisis, the progressive weakening of state institutions aggravated this situation, particularly among institutions dedicated to environmental matters.

Argentine officials hoped that the adoption of a target and the possibility of participating in the Kyoto mechanisms would help create new sources of employment, attract and channel investment into diverse sectors, and produce spin-off benefits for the entire economy. Yet, the lack of a detailed, systematic evaluation of all potential effects of the voluntary commitments makes it very difficult to determine whether the target is consistent with Argentina's long-term interests or to evaluate the conditions under which such a commitment could benefit the country.

Most of the relevant actors agreed that the adoption of an emissions target was not a national priority. This was true at the time, and is even truer today, with the deepening of the economic recession and the financial crisis that has followed in Argentina. In the past 2 years this crisis has dramatically changed priorities in Argentina, compared to the 1990s. However, the actors consulted still believe it makes sense to gradually adopt policies and measures that naturally lead to limiting emissions, as long as doing so does not affect economic growth.

Generally in Argentina, there is little public awareness of environmental matters. This is true in political, technical, and business circles; in the media; and in particular among opinion-makers, as well as among the general public. Achieving the target would entail not only creating greater awareness in the population but also developing a political agenda and an environmental strategy on climate change. Meeting a commitment would require strengthening the institutions that foster participation, a problem of public policy in general, not only environmental policy. Although civil society should potentially play a fundamental role in defining policies related to climate change—such as elaborating mitigation proposals and turning these proposals into norms, standards, and planning—no procedures to assure its participation are in place.

The complexity and ever-evolving nature of the UNFCCC negotiation process make it difficult for sectors such as academia to systematically follow and monitor the negotiations, as they do not have the necessary financial support. The result is that the academic sector's relationship to this process is intermittent, which blocks the possibility of greater participation in the process.

Until the population's everyday subsistence problems are resolved, it will be very difficult to raise public awareness of climate change or improve public participation in the elaboration and definition of policies. Only then will the average citizen have the time and space to reflect on issues that do not seem urgent when compared with meeting basic daily needs.

Taking Advantage of Opportunities in an Eventual Emissions Market: The Tradeoff between Mechanisms and Targets

Future opportunities for non-Annex B countries have been discussed in meetings and in documents on the flexibility mechanisms and on the future markets for trading GHG emission credits. Although countries could benefit by taking advantage of the Kyoto mechanisms and taking "early action" on climate change, they differ in their capacity to influence emission reductions markets and thus benefit from the mechanisms.

In this sense the opening of a global market to trade emission credits would create opportunities for the large suppliers (e.g., China, India, and economies in transition). Together these countries could supply more than 80 percent of the demand. The remaining 20 percent could be distributed among 130 other countries, including Brazil, Mexico, Indonesia, Malaysia, and Korea.[22] It is also undeniable that "hot air" will serve to depress price levels, and consequently reduce the range of viable mitigation options for developing countries such as Argentina. Due to higher mitigation costs in Argentina, low market prices mean that Argentina would not be able to take advantage of opportunities to reduce emissions through the CDM, with the possible exception of carbon sinks. So-called no-regrets mitigation projects are one option for Argentina, although many believe that no-regrets options will be harder to get approval for under the CDM.[23] These options, however, would be available if Argentina had access to the rest of the Kyoto mechanisms.

Argentina would have to assume the costs for all actions resulting from voluntary commitments. Emission reductions in compliance with the voluntary commitment cannot be traded (as they are a part of the commitments assumed). The most rational approach to compliance would be to carry out the cheapest measures first, or those with a negative cost, and take advantage of the most favorable situations. Keeping in mind the "no regrets" concept, it would be worthwhile to voluntarily reduce only those emissions that imply assuming an incremental cost that is negative or zero. Yet, Argentina's voluntary commitment was so ambitious as to eliminate the possibility of generating surplus emission reductions. A lesser commitment would have meant a greater available surplus. The real benefits of making such a commitment stem from the possibility of trading the surpluses.

To be economically rational, the price anticipated from the eventual sale of offsets in the market should be higher than the cost of generating the emission reductions. This includes both the possibility of using no-regrets measures, as well as the possibility of taking other actions and enacting measures with positive costs (again, so long as these costs are less than the selling price of offsets in the market). The act of making a voluntary commitment should be based on an exhaustive evaluation of diverse possible scenarios and of the evolution over time of the variables involved to ensure that no economic losses result.

To evaluate whether a non-Annex I country should carry out early mitigation actions, the government should compare the opportunity cost of sacrificing the cheapest mitigation options today with the cost of having to carry out more expensive mitigation options in the future (as the result of commitments that may have to be made down the line). Obviously, these calculations involve high levels of uncertainty due to the behavior of diverse variables. They should be based on discounting the cost of future actions and comparing them with present mitigation actions that could be implemented.

IV. Final Considerations

Argentina's experience in proposing a target yields valuable lessons for others considering a voluntary target. A developing country contemplating making such a commitment should carefully and cautiously evaluate the target's type and level, as well as the conditions set for real compliance. The most salient considerations should include the following:

Need for thorough technical assessment. In the specific case of Argentina, the decision to adopt emissions targets was eminently political. More technical and economic studies were needed to support such a decision. The few studies done on Argentina's potential participation in an emissions market, and the magnitude of this market, did not provide serious arguments on which to base the adoption of the target. In Argentina's target evaluation exercise, resources and political will were needed to mobilize and enhance the existing capacity. Developing countries considering target taking should ensure they have adequate and proven methods, methodologies, and modeling capacity to ensure proper analysis.

Argentina's target proponents underlined the positive effects that "early action"—that is, target taking—would bring. Early action was expected to provide (1) access to more modern, cleaner technologies; (2) the possibility of qualifying for entry into more environmentally demanding markets; (3) development of the institutional capacity necessary to implement such initiatives; (4) greater knowledge of technologies; (5) accelerated development; and (6) a reduction in the costs of all these factors. However, these advantages or benefits were not quantified or estimated.

Energy-intensive industry flocked to Argentina in the 1980s, which could occur again in the future, considering the country's natural resource endowment. The transfer of energy-intensive manufacturing activities to developing countries is thus an important issue that should be included in an assessment. The assessment should also evaluate the relevance and applicability of emission reduction measures undertaken by other countries to its own domestic setting. The transferability of the measures may be limited as each country's circumstances are different. Countries should not assume that what works in one context would work in their own.

Need to carefully consider country-specific emissions conditions. Although target proponents were careful to select a type of target appropriate to Argentina's circumstances, the wisdom in volunteering a target at all, given Argentina's GDP and energy structure, may be questionable.

As a result of the institutional and regulatory reform of Argentina's energy system and the country's dependence on international markets, there is a close tie between the prices of energy and agricultural products and the evolution of the prices of these "commodities" in international markets. The uncertainty of future price scenarios and their impact on the internal activity level of the sectors with the greatest emissions creates an additional difficulty for a possible emission-limitation initiative. Even within a legal and regulatory framework, the volume of emissions depends, in Argentina's case, on exogenous factors that internal policies do not control. A low level of activity in the agricultural sector would imply that emissions grew at a lower rate than GDP, while a high level of activity in this sector would imply a rise that is more than proportional to GDP growth. Neither of these scenarios would reflect increased or decreased efficiency; they would be a function simply of GDP structure.

In the 1990s, the institutional and regulatory structures of Argentina's energy system were transformed. The resulting changes included alteration in behavior during the decision-making process (new actors), growth in economic efficiency with a subsequent impact on prices, the introduction of technologies (which, although more efficient, emit more than the technologies they replaced), and the substitution of primary energy sources. The impact of these changes has been an increase in GHG emissions as a result of greater energy consumption and greater penetration of natural gas in electricity generation, as gas has been substituted for non-emitting sources, such as hydropower and nuclear. Although these changes are recent, initial evidence suggests that emissions in key segments of the energy sector, such as electricity generation, will grow, increasing emissions intensity as well as the absolute level of energy consumption and thus impeding implementation of the proposed reductions target.

Need to be conservative in economic projections. Argentina's adoption of an emissions target was based on expectations of a more favorable future economic situation. Despite the nine emission projections undertaken, it is clear that, given the current economic collapse in Argentina, these projections were based on overly optimistic assumptions. Developing-country caution in making commitments in the short term and, particularly, within the context of volatile emerging markets and a lack of sustained economic development over prolonged periods, seems justified given the Argentine experience.

Need to leave room for growth. Argentina, like other developing countries, has development needs that could be sacrificed through the implementation of target taking. Energy consumption depends on the demand for services and for energy use related to greater comfort, mobility, communications, and entertainment, among others. In many developing countries, a high percentage of the population cannot meet even its basic energy needs, making it logical that growth in income will also mean a concomitant rise in energy consumption. As a result, increase in income per capita could mean a non-linear increase in energy consumption per capita. A proposal to reduce emission intensity, therefore, could become an impediment to improving the population's welfare since this might translate into limits on the ability to satisfy energy needs.

For target taking, analysis should have taken into account the structure of Argentina's GDP and its vulnerability to significant cyclical shifts. Moreover, the industrialization process, even when based on efficient technologies, implies growth in emission intensity, as industry makes up an increasing share of economic production relative to other, less energy-intensive, sectors. Over the past 15 years, Argentina has gone through a process of deindustrialization that must be reversed in order to address the country's main economic and social problems (including unemployment, poverty, inequity, trade deficit, and external debt). The recovery of industrial growth is likely to increase emissions. This example is relevant to many developing countries that, like Argentina, are experiencing economic deterioration, rising poverty, low GDP levels, and increasing inequity. The reversal of this process could have an impact on emissions intensity.

Need for realistic assessment of technology and emissions-related markets. In a developing country such as Argentina, the greatest potential for improving efficiency depends on incorporating technologies developed by industrialized countries, which are delaying compliance with their emission reduction commitments, and thus slowing the pace of technological development. Stable, or even lower, fuel prices are anticipated, which do not provide an incentive for the development and incorporation of more efficient technologies, particularly if they require significant additional investments.

Need for domestic policy relevance and buy-in. It is not enough to make voluntary commitments within the context of international pressure, particularly when those commitments imply actions and policies affecting all productive sectors and civil society. The absence of a broad and democratic debate and the lack of support from different social actors for the target demonstrate that its implementation would face numerous obstacles. At best, the commitment will become a voluntary posture that will never really be put into action.

While it is true that the social, economic, and environmental dimensions are interconnected, in reality issues such as short-term material and social welfare are a much higher priority for civil society. In Argentina, members of civil society (with the exception of environmental interest groups and academics) have hardly participated in discussions on environmental issues, and representatives of productive sectors and the government in general have also shown little interest. This clearly shows that the environment is the concern of a small group of actors with a certain level of knowledge and understanding of the climate problem's importance, rather than a national priority. Environmental issues are becoming even less of a priority in the wake of country's deepening economic and social crisis.

Although the effort to include broader government and public participation in the target-taking exercise was insufficient, it represents a first attempt toward gathering disparate sectors and actors in such an exercise. The process clarified the need to set a national long-term strategy combining climate concerns and economic development. As such, the effort in Argentina was worthwhile and unusual for a developing country.

Need for long-term view. The selected actions also need to fit within the framework of long-term state policies. Otherwise a change in administration will yield policy changes that do not provide continuity. Argentina's announcement was made in the context of short-term politics and was not linked to longer-term sustainable development policies. The best evidence of this is the lack of agreement, even among state entities responsible for relevant sectors. This reality is clearly evident in the lack of concurrence and support from the energy sector.

Need for realistic understanding of international order. It is important to reemphasize that the administration, at the time it made the commitment, tried to portray Argentina as a country that shared the reality of industrialized countries, rather than emphasizing its status as a developing country. Other actions, such as applying for entry into the OECD and rumors of incorporation into Annex I, all reinforced this discourse. Perhaps this posture was based on the belief that the country was an important international player.[24]

As regards international negotiations on climate change, neither the Convention nor the Kyoto Protocol contemplates this type of commitment. This presents serious obstacles and a significant negotiation challenge, as any modification of the Kyoto Protocol is perceived as threatening by many developing countries concerned that change would lead to more unfavorable conditions for them. While searching for answers to a global problem, everything seems to indicate that it is advisable to stay within the framework of agreements and consensuses reached and to avoid creating conflictive situations. Many observers viewed Argentina's proposal as an attempt to destroy the consensus inside the G-77 and China or, at least, as more of a confrontational position than a conciliatory one. In addition, Argentina's close alliance with the United States entailed serious problems in its relationships with its natural allies, such as the main trading partners in the MERCOSUR regional integration agreement.

The national circumstances of each Party imply that proposals and formulas cannot easily be adapted to each country's situation. Perhaps just certain key elements could be identified that should be included in the design of any voluntary commitment. The adoption of different targets that take into account the different national circumstances of each country could make the participation of developing countries easier, even if it makes the monitoring and verification of the commitments more complex. It could be more attractive for some developing countries to make voluntary commitments in order to take advantage of certain business opportunities resulting from the Kyoto Protocol, as long as these can be positively linked to each country's development strategy.

Some authors maintain that the principal error in Argentina's proposal is that it rests on trying to gain access to mechanisms reserved for the Annex I/Annex B countries, while also retaining access to the CDM. This would imply opening up discussions on the Kyoto Protocol's modification or the possibility of creating other instruments that other key developing countries would not support.

Finally, the lack of serious initiatives from Annex I countries and donors is the context in which Argentina's proposal and proposals from other developing countries should be analyzed. The actions of these actors have not encouraged developing countries to participate in the process. The pressure of certain Annex I countries on developing countries to commit seems paradoxical: Those that are pressuring have already agreed to obligatory quantified targets, and yet many are not taking concrete actions to comply. Thus, for a developing country to seriously consider taking on a target, industrialized countries must overcome their resistance to implementing the reductions needed to meet the commitments to which they have agreed.

The need for a serious and viable plan to integrate developing countries into the quest for climatic sustainability is a permanent and open challenge. It will not be resolved through isolated formulas, indices, or commitments that are difficult to verify and monitor. Such measures would lack the necessary international consensus and, thus, would have dubious impact on solving this problem.

In the end, the Argentina proposal was inconsistent, unworkable, and lacked policy coherence and domestic support. It provides, however, an excellent case study about how a developing country—with its own mix of political, economic, social, and environmental forces—grappled with the complex challenge of non-Annex I states to participate meaningfully in the Kyoto Protocol. So conceived, it may begin to serve as a repository for "how to" and, equally important, "how not to" undertake international and domestic policy discussions about the adequate articulation to the global effort to prevent climate change.

Notes

1. Dr. Carlos Saúl Menem was President of Argentina from July 1989 though December 1995 and from December 1995 through December 1999 (second period). During his entire mandate, Engineer Maria Julia Alsogaray (President of COP 4 in Buenos Aires) was in charge of the Secretariat of Natural Resources and Sustainable Development (government environmental agency in Argentina). This secretariat is a ministry and is directly dependent on the executive branch. Alsogaray was in charge of all environmental matters and had a very close relationship with the President, receiving preferential treatment.
2. New Zealand also presented a proposal at an opportune moment, asking for commitments from all Parties. During COP 4, developing countries had heated

discussions with the United States, when it tried to force them into making voluntary commitments.

3. Even before the Kyoto Protocol, U.S. industries launched a campaign to convince the public that a strong agreement on climate change would increase prices of all goods and that countries such as China, India, Mexico, and South Korea would become "free riders," while U.S. industry would lose its competitive edge.
4. On December 10, 1997, when celebrating the agreement reached in Kyoto, the U.S. President stated that important challenges remained, particularly in ensuring meaningful participation of developing countries in mitigation efforts. He said that industrialized countries had already taken firm steps in terms of emissions mitigation by accepting quantitative commitments, noting that, "We do not accept binding obligations unless key developing countries meaningfully participate in this effort." On August 12, 1997, then-Vice President Al Gore reiterated that in order to ratify the treaty and send it to the Senate, developing countries would have to participate in a significant manner. The problem is that "meaningful participation" is a vague concept that has not been defined, allowing the U.S. to reject any efforts by arguing that they are insignificant.
5. Declaration of Intention of the Governments of the Republic of Argentina and the United States of America on Cooperation for Sustainable Development, Joint Implementation and Jointly Implemented Activities to Reduce Greenhouse Gas Emissions. Signed on October 16, 1997 in the city of San Carlos, Bariloche by Guido Di Tella (then Argentine Foreign Minister) and Madeleine Albright. *Diario Clarín*, October 17, 1997, Buenos Aires; *Diario Río Negro*, October 17, 1997, Viedma, Río Negro, Argentina. It should be kept in mind that this declaration was signed before COP 3 and the establishment of the Kyoto Protocol.
6. The Bariloche Presidential Declaration. *Diario Río Negro*, October 17, 1997.
7. Argentina explored this alternative and led a consultation group made up of Latin American countries during the seventh meeting of the Ad Hoc Group on the Berlin Mandate (AGBM7). However, these consultations on a new proposal did not produce any concrete results.
8. The president of COP 4 was Engineer María Julia Alsogaray, who was also the Secretary of Natural Resources and Sustainable Development of Argentina.
9. See CSE (1998). As an example, the Brazilian delegation declared that it would not accept the inclusion of voluntary commitments on the agenda of the COP 4 meeting, and G-77 and China both supported this argument. These countries expressed their views in the conference with the Indonesian ambassador (chair of the G-77 and China at that time). See Ministry of Science and Technology (1999), http://www.mct.gov.br/clima/ingles/negoc/interv6.htm. The Brazilian position was ratified during various public presentations by its delegations. For example, see United Nations (1998), http://www.un.int/brazil/speech/98d-2com-clima.htm. It is important to consider Brazil's position on this matter and to keep in mind that it is Argentina's main trading partner in MERCOSUR, along with Uruguay and Paraguay, and that the productive structures of both countries are integrated (particularly the automotive sector).

10. It should be recognized that María Julia Alsogaray's intention was extemporaneous, since this matter had already been decided in 1997 and was revisited just before COP 4. It was not politically feasible to present this issue once again in November 1998.
11. Castellini (1998). It was argued that the creation of commitments was not the original intention and that these commitments were not meant to be obligatory. The idea was to permit countries to do so if they wished, and voluntary commitments were intended just to limit emissions, not eliminate emissions. However, the proposal was not even discussed.
12. See Gobierno de la República Argentina (1999) concerning the Argentine situation, and Zhang (2000, 2001), referring to the global situation.
13. The actors consulted include individuals from academia, NGOs, business, government, and interest groups, as well as the officials responsible for the design and development of the target.
14. Diverse military actions, such as the immediate participation with peace-keeping forces and support for the Gulf War (sending of military aircraft), as well as votes in various international forums, are all notorious examples of this alliance.
15. In addition to the government bodies mentioned in the text, the Commission included the Secretary of Economic and Regional Programming; the Secretary of Industry, Commerce, and Mining; and the Secretary of Transport, all within the Ministry of the Economy and Public Works and Services.
16. The level of agricultural activity is strongly influenced by the international meat and grain market. As a result, emissions produced by this activity do not depend on the behavior of local economic variables but on the expected international context for future prices of primary products.
17. As Table 6.1 shows, the method of the square root target in all the cases (scenarios) guarantees a positive emission reduction, both in absolute value and in percentage, in comparison with the BAU scenario.
18. Barros and Conte Grand (1999).
19. In this case, the emissions permitted during the commitment period total 100.4 million tons of carbon equivalent.
20. This issue was analyzed in Gobierno de la República Argentina (1999), based on a paper by Embree and Wilkinson (1999). Also see Girardin (2000).
21. This is the main reason for the insistence on maintaining access to the CDM.
22. MIT (1997), cited in Gobierno de la República Argentina (1999) and Girardin (2000).
23. Although there are many documents on this topic, we recommend Reid and Goldemberg (1999).
24. Another telling example is the pretension to mediate in the Middle East conflict.

7. The Brazilian Proposal on Relative Responsibility for Global Warming

Emilio L. La Rovere, Laura Valente de Macedo, and Kevin A. Baumert

Introduction

This chapter examines the policy implications and future potential of the Brazilian Proposal for establishing limitations on greenhouse gas emissions. Considerable confusion surrounds this Proposal, in part because the term "Brazilian Proposal" has two meanings. One refers to the specific proposal introduced in the United Nations Framework Convention on Climate Change (UNFCCC) negotiations prior to the 1997 adoption of the Kyoto Protocol (UNFCCC 1997b). Like other proposals put forward during Kyoto negotiations, the Brazilian Proposal offered a concrete option for structuring the Protocol's emission limitation requirements. The other meaning refers to elements of the Brazilian Proposal that have persisted since the Protocol's adoption. The most notable of these is the burden-sharing scheme, which apportions greenhouse gas (GHG) emissions targets according to each country's historical responsibility for the global temperature increase. As such, the burden-sharing scheme can be understood as a general methodological framework for determining emission limitation commitments among states. This aspect of the Brazilian Proposal is very much still alive; research and analysis continues (MCT 2000; UNFCCC 2001).

This chapter has three sections. Section I describes the basic features of the original Brazilian Proposal as presented in the Kyoto negotiations. It also explains the proposal's political significance in the negotiations. Section II discusses the Brazilian Proposal's methodology in a policy context, including various ways of determining historical responsibility. Section III analyzes the future implications of adopting the proposal's methodology under the Climate Convention and, in particular, the pros and cons of using its burden-sharing scheme to establish emission targets for the Kyoto Protocol's second commitment period and beyond. This section suggests some modifications to the Brazilian Proposal that could make it more feasible in the future.

I. The Context and Features of the Original Brazilian Proposal

In 1995, the first Conference of the Parties (COP 1) to the Climate Convention adopted the Berlin Mandate, which stated the need to establish GHG emissions reduction targets for industrialized and transition (Annex I) countries and affirmed the implementation of the Convention's commitments for developing (non-Annex I) countries. The negotiation process aimed to adopt a protocol to the Climate Convention at COP 3, in Kyoto, Japan, in 1997. In the run-up to Kyoto, the process invited proposals from all Parties on how to shape "quantified emission limitation and reduction objectives" for Annex I Parties. Many proposals included indicators—or a combination of indicators—such as per capita emissions, gross domestic product, and energy intensity, among others (UNFCCC 2000c). The idea of a Brazilian proposal was developed between 1996 and 1997 by experts from the government and the national scientific community and, particularly, by Luiz Gylvan Meira Filho and José Domingos Miguez from Brazil's Ministry of Science and Technology, in consultation with Luiz Pinguelli Rosa, professor at COPPE/UFRJ.[1] In July 1997, the Brazilian government presented its proposal to base emission reduction requirements on an industrialized country's relative responsibility for the global temperature rise (currently about 0.6° C higher than pre-industrial levels).

Specifically, the Brazilian Proposal called on Annex I countries as a bloc to reduce their GHG emissions 30 percent below 1990 levels by the year 2020. The proposed reduction target covered the three main GHGs (carbon dioxide, methane, and nitrous oxide) and extended from 2001 to 2020 (using a succession of 5-year commitment periods).

The Proposal's most innovative feature was the method used to distribute emission reduction burdens among countries—according to each country's relative responsibility for the global temperature increase. The Proposal included a complex methodology for determining this responsibility for individual Annex I countries, as well as for determining the associated targets (the target methodology was *not* applied to developing countries in the Brazilian Proposal). It also suggested the need for an "agreed climate-change model" for estimating each country's contribution to global temperature increase and, as an illustration, included a "policymaker model" for estimating country targets (UNFCCC 1997b).

One consequence of the Brazilian Proposal's approach and methodology was that countries that industrialized earlier tended to incur the largest emission reduction requirements in percentage terms. For example, in the original Brazilian Proposal, the indicative target for the United Kingdom was a 66 percent reduction below 1990 levels by 2010, while the targets of the United States and Japan were about 23 and 8 percent, respectively, reflecting the fact that they industrialized more recently than the United Kingdom (UNFCCC 1997b). However, these targets and others presented in the original Proposal were illustrative only and were later shown to have some methodological shortcomings (see Section II). The Proposal incorporated flexibility into targets by allowing individual targets to be negotiated among Annex I countries. In other words, Annex I Parties would be bound by the collective target and could trade individual targets among themselves.

Another important element of the proposal was the Clean Development Fund (CDF)—a punitive and financial mechanism to be managed by the Global Environment Facility. Failure on the part of industrialized countries to achieve their required reductions would result in a fine, payable to the CDF. The value attributed to the fine was set at US$10 per ton of carbon emissions exceeding the target. The distributive criterion for the fund corresponded to the Brazilian Proposal's rationale of proportionality: Non-Annex I countries could apply for funds according to their relative contributions to atmospheric warming. The funds would finance GHG abatement projects, and up to 10 percent would be used for adaptation projects. The primary objective of the proposed fund was to promote climate protection, including through the transfer of clean technologies and allowing for the participation of non-Annex I Parties.

With these features, the Brazilian Proposal addressed two key issues that pre-Kyoto negotiations were attempting to address (UNFCCC 1997b, 9). First, it addressed the issue of "the future level of emissions to be tolerated from the Annex I Parties" (i.e., the "cap"). Second, the Proposal suggested a "criterion for the sharing of the burden" among industrialized countries (i.e., by historical responsibility for temperature increase). The Proposal also dealt with the issue of developing-country participation in a manner consistent with the Berlin Mandate (which called on industrialized countries to take the first quantitative commitments). Accordingly, the Brazilian Proposal did not call for developing-country commitments, but rather for developing countries to share the CDF's proceeds in order to implement "clean development" projects.

The COP in Kyoto did not adopt the Proposal. Industrialized countries considered the methodology for estimating past emissions to be biased (see Section II)[2] and insisted on negotiating targets in a bottom-up, pledge-based fashion. Some industrialized-country Parties felt that the Brazilian methodology unfairly punished countries for actions in the past, when the consequences of emitting GHGs were unknown.

The CDF was rejected because, among other reasons, it was a punitive instrument entailing financial penalties, making it an unlikely instrument in an international treaty. The allocation of resources from the CDF to developing countries was also considered questionable. As noted above, the resources would be distributed to non-Annex I countries in proportion to their relative contribution to global temperature increase. In other words, the higher their contribution, the more resources they would receive. According to the simulation undertaken in the Proposal, China would receive the largest share of the funds, about 32 percent (UNFCCC 1997b). In light of these shortcomings, the Kyoto Protocol negotiations subsequently modified the CDF into what is now known as the Clean Development Mechanism (CDM), which earned widespread support from industrialized and developing countries alike.

Over and above the CDM, the Brazilian Proposal played a significant role in the Kyoto negotiations, and it should be understood within that historical context. Despite the fact that the 1995 Berlin Mandate called for quantitative commitments for industrialized countries only, the United States and some other developed countries were pressing hard in the Kyoto negotiations for such commitments from developing countries. The Brazilian Proposal helped defuse the arguments posed by the United States and others that developing countries should adopt emission limits. In a highly political debate, the Proposal used scientific considerations, the well-established polluter-pays principle, and the Climate Convention principle of responsibility to argue, if implicitly, against developing-country commitments.

Moreover, the Proposal helped to further engage developing countries in the debate over the emission commitments of Annex I. The analysis included in the Proposal illustrates that developing countries have an important stake in precedent-setting quantitative commitments adopted by Annex I (UNFCCC 1997b). Given the imbalances in negotiating power between industrialized and developing countries, the Brazilian Proposal framework held the potential to inject transparency and objectivity in target setting, which might improve the future likelihood of developing countries receiving fair treatment.

II. Defining Responsibility for Climate Change

The Brazilian Proposal is more than a proposal presented in the Kyoto negotiations, it is a framework for allocating emission reduction burdens across countries and a subject of continued debate and analysis. After COP 3, the Brazilian Proposal was referred for further methodological analysis to the Subsidiary Body for Scientific and Technological Advice (SBSTA). The SBSTA review includes an effort by the Secretariat of the UNFCCC to promote debate and information sharing.[3] The Secretariat has organized expert meetings to review the Proposal, aiming to "identify issues relating to the scientific and methodological aspects of the Brazilian proposal, including those that need further consideration and areas of future work" (UNFCCC 2001). Although this assessment does not directly address policy implications, the Brazilian Proposal is currently the only such proposal being officially considered by the Parties.

Two policy-related questions are important to the ongoing review and study of the Brazilian Proposal methodology. First, how well does the Brazilian Proposal methodology capture the relative contributions to warming? Second, are relative contributions to warming the appropriate measure of country "responsibility?" The remainder of this section examines these two issues in detail.

Methodology of the Brazilian Proposal

In response to scientific and technical concerns raised by experts, the Brazilian government revised the calculation method that accompanied its original proposal in 1999. The latest calculation methods for the Proposal (dated January 2000) are available on the websites of the Brazilian Ministry of Science and Technology and the Climate Convention Secretariat.[4] In its review of the Proposal, SBSTA is systematically investigating scientific and methodological issues. A background paper by the UNFCCC (2001) Secretariat identifies a host of scientific and methodological considerations worthy of consideration.

The Brazilian Proposal has already been the subject of considerable scientific study.[5] A landmark report by the Dutch research institute RIVM (Elzen et al. 1999) reviewed both the original and revised versions of the Brazilian Proposal. The study found that the revised version was "a major improvement with respect to the original version but still contains a few

Table 7.1. Regional Contribution of Greenhouse Gas Emissions to 1990 Temperature Change, *percent of total*

Country or Region	CO_2 from Fossil Fuels Only	Methane, Nitrous Oxide, and CO_2 including CO_2 from Land-Use Change
Canada	2.3	2.0
United States	31.2	21.7
Western Europe	21.7	16.3
Eastern Europe	5.8	5.0
Commonwealth of Independent States	14.8	11.9
Japan	4.2	2.8
Latin America	4.3	10.9
Africa	2.5	5.7
Western Asia	1.8	2.6
India	1.9	6.9
China	7.0	10.8
Oceania	1.2	1.7
Annex I	**81.2**	**61.1**
Non-Annex I	**18.8**	**38.9**

Source: Elzen et al. 1999.
Note: Results are derived from the EDGAR-HYDE data set and the meta-IMAGE model, not the Brazilian policymaker model.

shortcomings. The revised model still ignores the terrestrial part of the carbon cycle, and only focuses on the slow (oceanic) carbon dynamics" and contains some other characteristics that "seem to differ from those of other climate models." The overall effect, according to the RIVM study, is "an overestimation of the contribution of Annex I countries to temperature increase. These deficiencies can all be improved by corrections or by importing techniques and processes already available in other models."

In this regard, Elzen and colleagues analyzed the Brazilian Proposal with respect to the sensitivity of incorporating various GHGs and sources. Because of limited data availability, the original Brazilian Proposal considered only CO_2 emissions from fossil fuel sources. Table 7.1 shows that the relative responsibilities of countries and regions can change significantly when all sources of CO_2 (including from land use changes) as well as two other GHGs (methane and nitrous oxide) are included. The incorporation of all sources of CO_2, methane, and nitrous oxide reduces the collective responsibility of Annex I countries for temperature increase from 81 to 61 percent. Scientific and model uncertainties can also strongly influence relative responsibilities (Elzen and Scheaffer 2002). Interestingly, however, Elzen and Schaeffer show that if the relative responsibilities are calculated for *Annex I countries only* (consistent with the original Brazilian Proposal), the results are remarkably insensitive to modeling uncertainties and the inclusion of gases and sources beyond CO_2 from fossil fuels. This is partly due to the fact that developing countries have a much larger share of CO_2 emissions from land use changes and non-CO_2 emissions than from fossil fuel-related CO_2 (where industrialized countries dominate).

Indicators of Responsibility

Conceptually, the Brazilian Proposal is built on the "common but differentiated responsibilities" and "polluter pays" principles. These are important principles enshrined in the 1992 UNFCCC. Although these principles are widely accepted, the Brazilian Proposal is not without controversy. Responsibility is a normative concept with competing viewpoints.

To illustrate, Figure 7.1 shows a representation of the chain of causality linking emissions to climate change and impact (Enting and Law 2002). Each stage of the chain involves some degree of *delay*; thus, the farther down the chain one goes, the greater the delay between actions and their effect. Indicators of responsibility could be considered at different points along this chain. Locating responsibility farther down the chain (e.g., sea-level rise) will differ considerably from locating it on the top part of the chain (e.g., emissions). It is important to realize that industrialized countries will show a larger share of responsibility using indicators late in the chain—such as temperature change and sea-level rise—mainly because of the longer average "age" of their emissions in the atmosphere. Countries with more recent emissions will therefore show smaller shares of responsibility late in the chain because their

Figure 7.1. A Chain of Causality in Global Warming

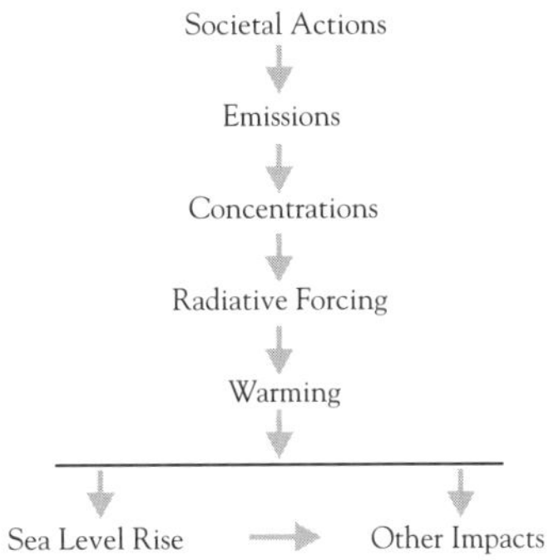

Source: Enting and Law (2002).

Figure 7.2. Temporal Relationship between Emissions, Concentrations, and Temperature Change

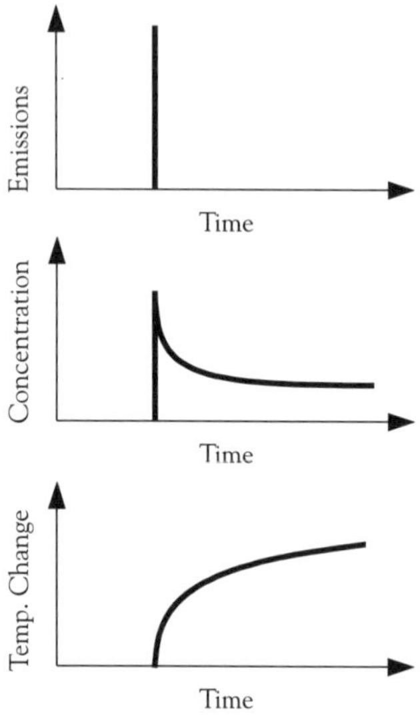

Source: Adapted from Höhne and Harnisch (2002); Elzen et al. (1999).

emissions have quite some time before reaching their full global warming potential.

These time delays are shown in Figure 7.2. The top frame shows a hypothetical "pulse" of emissions at a given point in time (with *no* emission before or after this pulse). This emission pulse leads to an immediate increase in atmospheric concentrations (middle frame), which declines over time as the gas is slowly removed from the atmosphere (e.g., through decay out of the atmosphere). Finally, the temperature change (bottom frame) resulting from the pulse continues into the future, even as the concentration is declining. Understanding these delays is essential to understanding the Brazilian Proposal and how it relates to other indicators of responsibility.

Time lags explain why the Brazilian Proposal has such different consequences than measuring responsibility with emissions or concentrations. To illustrate, Table 7.2 shows the relative responsibilities of Annex I and non-Annex I countries using three different indicators. The first is *annual emissions* (1990) and the second is contributions to increases in CO_2 *concentrations*, which is a function of historical emissions over time. Finally, the contribution to *temperature increase* (i.e., actual warming) is shown for 1990 and projected for 2010 and 2020.[6] Each of these indicators shows that industrialized countries are primarily responsible for climate change. However, industrialized country responsibility is largest when expressed in terms of temperature increase, due to the long atmospheric residence time of CO_2 and the past warming influence of CO_2 that is no longer in the atmosphere. Even though most emissions from 100 or more years ago are no longer present, their influence on global temperature lingers.

Under these different indicators of responsibility, the date at which developing- and industrialized-country responsibilities reach parity varies dra-

Table 7.2. Relative Contributions of Annex I and Non-Annex I Countries to Global Climate Change

Indicator of Contribution	Percentage Shares	
	Annex I	**Non-Annex I**
Emissions in 1990	75	25
Concentrations in 1990	79	21
Temperature increase:		
in 1990	88	12
in 2010	82	18
in 2020	79	21

Source: Adapted from Pinguelli Rosa et al. (2001).
Note: Includes only CO_2 from the energy sector.

matically. Using *annual* emissions, the IS92a scenario of the Intergovernmental Panel on Climate Change (IPCC) suggests that developing and industrialized countries will reach parity at about 2037. Yet, under the original Brazilian Proposal methodology, Annex I and non-Annex I responsibilities reach parity in 2147[7]—a delay of more than 100 years compared to calculations based on annual emissions. More recent estimates using different models and data sets suggest different dates. Analysis by Elzen and Schaeffer (2002) suggests convergence of Annex I and non-Annex I contributions in 2015 for CO_2 emissions, 2045 for CO_2 concentrations, and 2055 for temperature increase. Austin et al. (1998) estimate annual emissions parity at 2015 and concentration parity at 2057 (or 2038 if CO_2 emissions from land use are included). As explained in the previous section, if emissions from land use changes (e.g., deforestation) and all gases are included in the analysis, the date at which industrialized and developing countries reach parity under any indicator moves closer to the present.

Proponents of the Brazilian Proposal (UNFCCC 1997b), however, argue adamantly against formulating responsibility using annual emissions:

> *It is often implied that...most of the responsibility for climate change in the future will tend to be attributed to non-Annex I Parties, the year when the non-Annex I emissions equals those of Annex I Parties being taken as the year when the respective responsibilities become equal. This approach for implicit differentiation of responsibilities overestimates the non-Annex I Parties share of responsibility, as it does not take into consideration the different historical emission path resulting from very different industrialization process and consumption patterns in time of both groups.*

Because climate change is caused not by *emissions* but by the rising *concentration* of GHGs in the atmosphere over time, the argument against using annual emissions as an indicator of responsibility is a strong one.

Nevertheless, the Brazilian Proposal's representation of responsibility is not without controversy. The Proposal takes into account only the temperature increase that has *already occurred*. It does not consider the extent of future warming that the present increase in atmospheric concentrations has committed us to. For this reason, the Brazilian Proposal weights past emissions significantly more heavily than emissions in recent years. Yet, recent emissions will undoubtedly have an effect on future warming. In this regard, Höhne and Harnisch (2002) suggest than an appropriate indicator for responsibility should be not only *backward*-looking (such as the Brazilian Proposal) but also *forward*-looking. Similarly, according to Elzen et al. (1999), "It might make sense to include some form of 'forward-looking' assessment in the analysis of countries' responsibility for global mean temperature increase. In such an approach, not only would the current effect be evaluated [as in the Brazilian Proposal], but also the future effect of greenhouse gases emitted in the present and the past."

III. The Future Potential of the Brazilian Proposal: The Burden-Sharing Scheme

This section analyzes the future implications of possibly adopting principles of the Brazilian Proposal within the UNFCCC framework, with particular attention as to whether its burden-sharing scheme could be used to set emission targets in subsequent rounds of the negotiation process for the Kyoto Protocol.

Continued Validity and Usefulness

Like most proposals made before the Kyoto Protocol's adoption, some parts of the original Brazilian Proposal are no longer applicable to the current negotiations. For example, the adoption of the CDM and the non-compliance procedures (through the 2001 Marrakesh Accords) suggest that the CDF will have little applicability in the future negotiations.

The Brazilian Proposal's burden-sharing scheme, however, continues to be a useful idea and could offer an approach to bring non-Annex I countries aboard the emission control system. So far, negotiations on developing-country emission limitations under the Climate Convention have been deadlocked. Developing countries insist on establishing a connection between the Climate Convention goals and sustainable development through mechanisms that transfer financial resources and technology from North to South. In contrast, Annex I countries focus on their economic losses due to mitigation of GHG emissions and emphasize the need for developing countries to come aboard to achieve the Convention's objective of preventing dangerous climate change. The Brazilian Proposal supplies a starting point to break this deadlock. While focusing on the main goal of stabilizing the global climate, it quantifies the different individual contributions of each Party to the existing global temperature increase and, consequently, to the required efforts to solve or minimize the problem.

The proposed approach is science-driven. This is good news, as it avoids a burden-sharing scheme based solely on the bargaining power of Parties sitting at the negotiations table. Arrangements driven by sheer negotiating power are subject to all kinds of asymmetries and imperfections, as illustrated in the process leading to the establishment of Kyoto targets. However, it should be noted that the Brazilian Proposal's approach to establishing responsibility is not free of dissension, as discussed above.

Finally, the adoption of the Brazilian Proposal's burden-sharing scheme would be compatible with the Kyoto flexibility mechanisms, allowing for the deployment of market forces to help the scheme become easily operational. In fact, the Brazilian Proposal mainly addresses the establishment of targets for limiting the emissions of the UNFCCC Parties (i.e., the burden-sharing scheme) and can leave delicate implementation issues, such as compliance and limits to emissions trading, open for future negotiations.

Main Difficulties of Application and Possible Adaptations

The Brazilian Proposal's burden-sharing scheme faces some major difficulties that hamper its capacity to be immediately operational in the Climate Convention negotiations. These obstacles are discussed here, together with some possible adaptations that could improve the overall feasibility of adopting the general principle of the Brazilian Proposal in the next Kyoto rounds.

Complexity

As pointed out by Depledge (Chapter 2), "complexity can kill even the most intellectually brilliant proposal." This is a challenge for the Brazilian Proposal's burden-sharing methodology, which would require an agreement that incorporates complex scientific models and other technical considerations. One way to simplify would be to use the *cumulative* GHG emissions of individual countries from some given year in the past (to be negotiated). This would avoid the need to use, and agree on, a particular climate model. The scientific rationale for doing so is based on the literature findings and recent IPCC work (IPCC 2000a), which has shown that cumulative emissions supply a reasonable "proxy" for the relative contribution to global warming of different Parties to the Climate Convention, when considered in a time period limited to a few decades. Actually, the Annex I/non-Annex I crossover dates for GHG concentration in the atmosphere (a function of cumulative emissions) and temperature increase are relatively close (e.g., only 10 years apart, 2045 and 2055, according to Elzen and Schaeffer (2002)).

Outreach and education efforts could also be used to overcome the complexity barrier. A systematic strategy is needed to better explain the methodology in terms accessible to wider audiences. This would include building on sparsely available previous attempts (e.g., La Rovere 1998). Through workshops and materials, the strategy would then disseminate the "user-friendly" information via an outreach campaign to stimulate public education and international awareness in a few target audiences in particular.

Data

Going back to the 19th Century presents serious problems because of the need for reliable GHG emissions data from individual countries to serve as a basis for negotiating future targets. CO_2 emissions from fossil fuels would be the least controversial data set to agree on. Elzen et al. (1999) show Brazilian Proposal calculations using three different fossil-fuel CO_2 data sets. The resulting sensitivity analysis shows relatively small differences in results. Even so, it may be difficult to reach a consensus on figures for the distant past. Experience with in-depth reviews of emission inventories has shown a number of difficulties in estimating emissions even for recent years. (See, for example, UNFCCC 2000a.) Furthermore, the Kyoto Protocol negotiations illustrate the political need to base decisions on official data supplied by each government rather than on the estimates of international organizations or research institutes or on the worldwide estimates of single government agencies.

The data challenges would be greatest for CO_2 from land use change and emissions of non-CO_2 GHGs. Again, the inclusion of these sources and gases will have a considerable impact with respect to attributing responsibility for warming at the global level (Elzen and Schaeffer 2002). Here, it seems difficult to even reach an agreement on accurate data for current years (see IPCC 2000b), and insurmountable obstacles might arise in determining what figures to use for the 18th and 19th Centuries. One possible remedy to this difficulty would be to conduct extensive sensitivity analyses for different historical data sets of CO_2 from land use change and non-CO_2 gases. These analyses might identify the most significant discrepancies and enable analysts to develop estimates for use in the negotiations. Currently, however, very few historical data sets exist for these sources and gases at the country level. In any case, substantial improvement in the quality of land use change and forestry data, as well as non-CO_2 data, would be required at the national level.

Of course, data reliability problems would be reduced if the starting year for the accounting of cumulative emissions were established in the 20th Century: the closer to the present, the higher the quality of data. From 1990 on, this problem can be solved through proper review of inventories presented as part of national communications to the Climate Convention. However, the acceptability of such a late starting year for accounting of cumulative emissions remains to be proved at the negotiations table (see La Rovere 2002). Finding a balance to these data issues seems a daunting challenge.

Acceptance of responsibility for pre-1990 GHG emissions

The issue of responsibility for past emissions was first raised by a group of Indian scientists, under the leadership of the late Anil Agarwal, during preliminary discussions on the creation of the Climate Convention. Annex I countries then dismissed this argument, maintaining that they could not be blamed for their past GHG emissions' negative impact on world climate when they did not know about the consequences of burning fossil fuels at the time. According to this view, the first year to be taken into account would be 1990, when the IPCC published its First Assessment Report warning that GHG emissions could have been contributing to global warming. On the other hand, many countries have laws and regulations embracing the legal principle of "objective responsibility"; for example, in the United States and Brazil, a polluter cannot escape a penalty by claiming unawareness of the environmental damages caused.

Once again, a possible solution to this problem would be to use contributions to cumulative emissions from 1990 to the present, with a continuous update. Of course, this approach would favor Annex I countries, compared to the burden-sharing approach currently espoused in the Brazilian Proposal. Alternatively, a compromise could be reached if an earlier starting year were established, based upon previous warnings about the gravity of the climate change menace from the scientific community or government reports (e.g., the 1960s report to the U.S. government from a commission of scientists chaired by R. Revelle et al.).

Bringing aboard developing countries

The Brazilian Proposal was originally devised as a burden-sharing scheme to be applied solely to Annex I countries. An important characteristic of the approach is that it yields emission targets in terms of absolute *reductions*. The Proposal does not currently allow for growth targets. This is problematic for any global application of the methodology. As outlined in Chapter 1, developing-country emissions will need to grow to meet economic development needs. Moreover, these countries' emissions are historically low and have contributed to climate change only in a small way. Imposing emission *reduction* targets on most (if not all) developing countries in the near and medium terms would be viewed as unfair and politically impossible.

One remedy might limit the application of the methodology to Annex I countries and shape commitments for developing countries on other bases (policies and measures, for example). Another possibility might be to use contribution to temperature increase or to global cumulative emissions as the starting point for negotiating targets for reducing emissions relative to a dynamic baseline (of the business-as-usual kind), rather than to a base year (this idea is illustrated in Chapter 9 of this volume). Additionally, individual or collective (non-Annex I countries as a whole) thresholds can be negotiated, below which countries would not need to commit to emission targets. Along these lines, Berk and Elzen (2001) examine a "participation threshold" based on income per capita.

Through these modifications to the Proposal, annual GHG emissions from Annex I countries as a whole would be required to decline continuously, while those from non-Annex I countries would be allowed to increase during an initial period, eventually stabilize, and finally decline until the end of the century. This kind of "safe-landing" analysis can build on the recent IPCC reference scenarios (IPCC 2000a) and the corresponding stabilization scenarios (IPCC 2001c). Informed by these scenarios, the duration of the grace period for non-Annex I countries to be free from mitigation targets would be negotiated.

Another option is to delay the participation of all developing countries until the relative responsibility of developing countries exceeds that of the industrialized countries. Prior to this date, the developing countries would have a grace period. The IPCC Special Report on Emission Scenarios (IPCC 2000a) estimated the dates when *cumulative* CO_2 emissions since 1800 from non-Annex I countries as a whole would overtake those from Annex I countries, according to different global reference scenarios. The results cover a wide range of possible pathways and outcomes, with the cross-over dates varying from the year 2040 (under the A1 scenario) to 2050 (A2 and B1 scenarios) and 2110 (B2 scenario). Similar analyses could be easily undertaken for cumulative GHG emissions since 1990, and the corresponding dates could be anticipated. When cross-over occurred, emission-reduction commitments could be established for non-Annex I countries on the basis of each country's relative contribution to cumulative GHG emissions since 1990. This approach could also provide an incentive to Annex I countries that are taking the lead, as the sooner they start implementing mitigation actions, the sooner non-Annex I countries will be brought aboard.

Alternative approaches are also possible. For example, before Annex I/non-Annex I cross-over occurs, some individual non-Annex I countries (those that really matter in terms of contribution to climate change) might reach a given threshold of relative responsibility. Such a threshold could mark the end of the individual grace period to which they are entitled. Once a country reaches such a threshold, it would then be required to take a mitigation target, provided that a corresponding financial compensation is established under the Convention. The specific level of such a threshold could be negotiated and settled according to different criteria. Again, the analysis of long-term global and national GHG emission scenarios would provide useful inputs to this discussion.

The Need for Further Research

Taking the Brazilian Proposal's burden-sharing methodology as a starting point, further research could explore long-term global GHG emissions scenarios to illustrate the combined effects of different trajectories of Annex I and non-Annex I GHG emissions. This analysis would supply useful insights to the negotiations on the initial date of non-Annex I countries' commitment to mitigation targets, according to different targets for long-term stabilization of GHG concentrations in the atmosphere. The comparative modeling effort sponsored by the UNFCCC already provides an appropriate framework to explore this research agenda.

IV. Summary and Conclusion

The Brazilian Proposal was a positive influence on Kyoto Protocol negotiations. Although it was not adopted, the Proposal continues to influence the debate over the contentious issue of developing-country commitments and the shape of what has become the CDM. Moreover, its burden-sharing principle—a core element of the original Proposal—is the subject of continuing review and study by experts under the direction of SBSTA.

This chapter suggests several adaptations to the Brazilian Proposal approach that might increase its acceptability and effectiveness. First, expressing responsibility in terms of *cumulative emissions* over time would reduce the need for complex scientific models and associated uncertainties. Complexity and uncertainty are likely to be major barriers to adoption. Second, governments might consider reducing the time frame during which responsibility is assessed. One option is to begin assessing responsibility in 1990, the date of the first IPCC Assessment Report. This could also address some of the political challenge of agreeing on data sets (especially non-CO_2 data and CO_2 from land use changes) from distant time periods that are not gathered or verified by governments. Third, to become operational on a global scale, an approach such as the Brazilian Proposal needs adapting to allow for growth targets. This need might be accommodated by shaping reduction commitments relative to a business-as-usual projection, rather than from a base year. Fourth, because many developing countries contribute little to global warming, it might be prudent to adopt a threshold for participation. For example, until a country reaches a certain level of responsibility or level of income, it would not be required to adopt emission limits.

Overall, these changes would preserve the original spirit of the Brazilian Proposal while making it more acceptable to Climate Convention Parties. These suggested changes would deliver a strong incentive to non-Annex I countries, which would be rewarded by any early action toward a lower-carbon development profile, as they would face milder mitigation targets in the future, no matter when such commitments came into force. Linking the end of non-Annex I Parties' grace periods to the emission reductions achieved by Annex I Parties could also provide an incentive for Annex I leadership.

Notes

1. Instituto Alberto Luiz Coimbra de Pós-Graduação e Pesquisa de Engenharia/ Universidade Federal do Rio de Janiero – Universidade do Brasil.
2. Bert Metz, personal communication, 2001.
3. The Secretariat has assembled a wide range of materials relating to scientific and methodological aspects of the Brazilian Proposal on the Internet (http://unfccc.int/ sessions/workshop/010528/documents.html).
4. The updated Proposal (Ministry of Science and Technology 2000) expands Appendix I of the original Proposal, adding new components to the formulas to reflect the climate system more accurately, but also retaining the simple version, the "policymaker model," as "the Brazilian Proposal." In addition, it includes a discussion of the concept of global warming potentials (UNFCCC 2001). Hard copies of the updated calculation methodologies may be obtained from Ministério da Ciência e Tecnologia, Gabinete do Ministro, Esplanada dos Ministérios, Bloco E - 3 Andar - Sala 398, 70067-900 Brasilia, Brazil.
5. See http://unfccc.int/sessions/workshop/010528/ for documents and other materials related to the Brazilian Proposal.
6. The scientific justification for the variables chosen in terms of projections and scenarios is based on the Intergovernmental Panel on Climate Change's (IPCC's) Second Assessment Report.
7. The original version of the Proposal (still located on the UNFCCC website as of late 2002) states that parity would be in 2162. After submitting the Proposal in May 1997, the authors realized that some of the calculations needed revision and asked the Secretariat to change this part of the document after the deadline. They sent the corrected version, but only the chart was included, not the table and the text with the new results. The new calculations demonstrate this parity date to be 2147 (José Domingos Miguez, personal communication, 2002).

8. EQUAL PER CAPITA ENTITLEMENTS: *A Key to Global Participation on Climate Change?*

Malik Amin Aslam

Introduction: The Kyoto Baggage

Ensuring broad participation among countries—including developing countries—is necessary to effectively address global climate change. Countries spanning the political divide acknowledge this, and almost all policy research carried out on the subject bears it out. Building on the principles agreed to under the 1992 United Nations Framework Convention on Climate Change (UNFCCC), the Kyoto Protocol stipulates a framework for the reduction of greenhouse gases (GHG) by the developed (Annex I) countries, while allowing the developing (non-Annex I) countries space to increase their emissions. At the same time, the Protocol charts a pathway for global cooperation via market-based regulatory instruments, especially international emissions trading (see Chapter 2 in this volume).

The Kyoto Protocol framework has two implications particularly relevant to any discussion of broad participation in climate mitigation. First, the Protocol uses in part the grandfathering principle by recognizing the 1990 emission levels of the developed countries as a basis for determining emission limitation targets. Grandfathering enabled national targets to be negotiated without any discussion of a long-term, environmentally sound, collective target. Second, no clear rule emerged to help differentiate targets *between* Annex I countries.[1] Targets were overwhelmingly shaped by the sheer force of bargaining power, exhibiting a system based on "negotiated justice" (Bierman 1999). However, what the Protocol fails to provide is a replicable framework based on any rationally defined criteria. This severely limits the ability of the Protocol to extend participation to developing countries in the future.

Current emissions of developing countries, as Chapter 1 points out, are very low compared with those of industrialized countries, but are rising rapidly. This places developing countries at a severe disadvantage when it comes to negotiating emission control targets that are based on a grandfathering system (which tends to establish targets in relation to a base year). Even in endeavoring to establish emission rights based on negotiated justice, the developing countries are handicapped by well-known global power imbalances: The developed world dominates the major controls on global capital, military prowess, and human capital. To the extent that the pure force of bargaining power has its way, the prospects that climate change negotiations will reach an equitable outcome are dubious. Thus, the Kyoto framework carries certain "baggage," which adversely bears upon many proposals for global participation.

The absence of an acceptable framework for expanding participation within the Kyoto Protocol—coupled with the necessity of involving developing countries for any future success of the climate regime—has compelled researchers to develop a variety of proposals attempting to expand participation while honoring the underlying Climate Convention principle of "differentiation" among countries. One of the most debated and controversial of the approaches focuses on equal per capita entitlements. A large bloc of developing countries steadfastly supports this approach.

Against this backdrop, the following sections describe the conceptual basis of an equal per capita entitlements approach and trace out the historical evolution of the idea. The analysis then gauges the strengths and weaknesses of the approach, as well its future applicability at the global level. Finally, the conclusion attempts to carve out a new proposal, a variant that attempts to keep per capita integrity while adding traits that neutralize or reverse traditional weaknesses.

I. Defining Per Capita Indicators and Entitlements

Generally speaking, "per capita" implies a number divided equally between a certain number of individuals. "Per capita" has two basic applications—as an indicator and as an entitlement. Most widely used, per capita is an *indicator* to represent, for example, the economic output or emissions of each individual in a particular country. As an *entitlement*, per capita is used as a measure to determine how much each country should be allowed to emit.

Per Capita Indicators

Dividing total economic output (i.e., gross domestic product), income, national debt, number of schools, and so on by the total national population generates indicators for use in comparing performance among countries. Such indicators have many uses; within the environmental sector, per capita indicators have been applied to the management of natural resources, such as freshwater and energy.

Within the climate regime, per capita indicators are used in various ways. First and most generally, per capita indicators help evaluate emissions disparities among countries. For example, the 2001 Marrakesh Accords state that industrialized countries "shall implement domestic action...with a view to reducing emissions in a manner conducive to narrowing per capita differences between developed and developing country Parties" (UNFCCC 2002). Second, per capita indicators are often used in policy debates relating to the *timing* of commitments to be taken by countries. For example, a per capita emissions indicator could be used to define a threshold for initiating developing-country participation into an emission control regime (much like the Montreal Protocol's staggered commitments for controlling ozone-depleting substances). Third, per capita emissions (and per capita GDP) are often proposed as an indicator for differentiating emission commitments between countries (Table 8.1). In brief, per capita indicators (emissions or GDP) can help facilitate an acceptable emissions-related burden-sharing agreement among countries.

As mentioned above, per capita indicators have already been incorporated into a number of proposals. Table 8.1 outlines proposals made by Parties during the Kyoto Protocol negotiation process[2] and evidences wide acceptance as well as use of this indicator in the negotiation process.

Although proposed per capita indicators as a "trigger" for participation and as part of burden-sharing schemes have remained relatively uncontroversial and uncontested, proposed "resource-sharing" schemes for shaping emission entitlements have stimulated intense debate and controversy, especially across the North-South divide. Under this approach, the global atmosphere—or, more precisely, the "limited assimilative capacity" of the Earth's atmosphere with respect to GHGs—is considered a global resource to which every human being is equally entitled. The remaining sections of this paper investigate this particular approach.

Table 8.1. Proposals from the Kyoto Protocol Negotiations

Proposal	Date of Proposal	Main Feature	Emission/ capita	GDP/capita
France	December 1996	Convergence	Y	
Switzerland	December 1996	Convergence	Y	
European Union	March 1997	Convergence	Y	
Norway	November 1996	Multi-criteria	Y	Y
Iceland	January 1997	Multi-criteria	Y	Y
Australia	January 1997	5 Indicators		Y
Japan I	December 1996	Indicator choice	Y	
Japan II	October 1997	Indicator choice	Y	Y
Poland	March 1997	4 Indicators	Y	Y
Estonia	March 1996	2 Indicators		Y
South Korea	February 1997	3 Indicators		Y

Source: Ringius et al. (2000).
Notes: Negotiations are from the Ad Hoc Group on the Berlin Mandate, prior to the Kyoto Protocol adoption. The table indicates that the per capita GDP indicator has been used as one of the indicators in some multi-criteria proposals.

Resource Sharing: Per Capita Entitlements

This approach first establishes an allowable level of global emissions, termed an emissions budget. The emissions budget (i.e., the total "environmental space," as Tynkkynen (2000) terms it) reflects the ultimate level at which to stabilize GHG concentrations over time, or the amount of GHGs that can be safely emitted in the atmosphere while meeting the ultimate objective of the UNFCCC.[3] This emissions budget is then distributed equally among the global population, thereby implying an equal right to the atmosphere, with each country getting an entitlement proportional to its population. These global budgets and the subsequent per capita entitlements can also be changed over time as new scientific information becomes available (Table 8.2).

Although there are some operational variants of this pure per capita approach,[4] this chapter focuses attention on the "convergence" scheme, which, in political and research circles, has become synonymous with this approach. In any case, all notable variants of this idea follow the generic approach outlined above.

The convergence scheme suggests that all countries participate in the emissions commitment scheme after the first commitment period of the Kyoto Protocol, with the ultimate objective of converging to equal per

Table 8.2. Emission Budgets and Per Capita Entitlements for Various Concentration Targets

Atmospheric Concentration Target (CO_2 parts per million by volume)	Emissions Budget 1991–2100 (millions of tons of carbon)	Average Annual Budget (millions of tons of carbon)	Per Capita Entitlement (tons of carbon)
350	300–430	2.7–3.9	0.5–0.7
450	630–650	5.7–5.9	1.00
550	870–890	7.9–8.1	1.3–1.4
650	1030–1190	10.3–10.8	1.7–1.8
750	1200–1300	10.9–11.8	1.8–2.0

Source: Adapted from Agarwal et al. (1999).

capita emissions over a stipulated time. As stated earlier, population size is a proposed criterion for determining how many entitlements each country is allocated. This scheme was first introduced by the nongovernmental Global Commons Institute (GCI) in 1990[5] and has been refined further into what is popularly termed "contraction and convergence." The approach has been consistently advocated at the sidelines of climate politics and, over the years, has received increasing support from some NGOs and governments.[6] However, to date, it has not been successful in breaking into mainstream climate negotiations.

GCI's approach starts off by defining a tolerable level of climate change based on the scientific assessments of the Intergovernmental Panel on Climate Change (IPCC),[7] which could be adjusted in the future to respond to improved scientific information. Based on such an ecologically sustainable target, a yearly global carbon budget is devised, which "contracts" gradually over time. This contraction continues toward a level where the per capita emission levels of participating countries "converge" toward an equal level. Thus, convergence claims to allocate shares of the budget to the emitting nations on an equitable basis (GCI 1999), whereby the per capita entitlements of the developed countries decrease while those of most developing countries increase. After reaching convergence, all countries would contract their emission entitlements equally until the requisite global emissions budget is reached. According to GCI, it is not possible to tackle the climate issue without adhering to these two key elements—contraction (environmental integrity) and convergence (equal per capita entitlements) (Meyer 2000).

The concept of a transition phase aims at softening impacts for Annex I countries. This staggered approach has been advocated by the Center for Science and Environment (CSE), an NGO, as a steady phase-in toward convergence (Agarwal et al. 1999). Gupta and Bhandari (1999), on the other hand, propose a scheme that differentiates between the short and the long term. While equal per capita entitlements are taken up as the long-term scheme, the approach is differentiated in the short term for Annex I countries by allowing them a transition period for adjustment. During this period, Kyoto Protocol commitments are fulfilled (until 2012), followed by targets in proportion to the efficiency of their production until 2025. After 2025, per capita entitlements are established. However, instead of aiming for an equal per capita convergence, Gupta and Bhandari suggest convergence toward a "sustainability corridor" (TERI 1997) of 0.5 to 0.75 tons of carbon per capita, which could accommodate the diversity of participating countries.

All proposed schemes, however, advocate the same underlying egalitarian concept of equal per capita entitlements for all human beings and seek acceptance of the principle within the context of the climate negotiation process. Before proceeding to analyze various aspects of this general approach, the next section briefly traces the evolving history of the entitlements approach.

II. History and Evolution of the Per Capita Entitlements Approach

The idea of equal per capita entitlements is older than the Climate Convention. In its treatment of equity and social considerations, the IPCC's Second Assessment Report (Banuri et al. 1996) cites Grubb (1989), Bertram (1992), Epstein and Gupta (1990), and Agarwal and Narain (1991) as some of the progenitors of the idea. Other early work on the idea includes Bertram et al. (1989), Smith and Ahuja (1990), and Smith et al. (1990).

Interest in equal per capita-based solutions intensified around the time of the initial report of the IPCC in 1990, which indicated the prospect of human interference with the global climate. This marked a watershed for the manner in which the atmosphere was viewed and signaled a rethinking of old paradigms. Whereas the view of the atmosphere as a global commons is not new, this report highlighted the atmosphere's finite assimilative capacity for tolerating GHGs from human activities. Exceeding the assimilative capacity of the atmosphere was one of the major factors influencing the accelerated changes in global climate patterns.

Thus, it was scientifically realized that there was a certain limit to the expansion of the global economy and attendant GHG emissions into the globally common atmosphere. This realization stimulated the associated debate on the *sharing* of scarce atmospheric resources, that is, establishing equitable access to this limited space, as the right to emit implies scarcity, and therefore economic value. As stated earlier, the suggestions for equal per capita entitlements were voiced. GCI and CSE continue to be the leaders on this front, although there are differences in their approaches.[8]

While the idea of equal per capita allocations was one of the core issues as negotiations began in 1991 to establish the Climate Convention, the time was not yet ripe for this approach. Convention negotiations shifted the focus toward differentiation of commitments within the *developed* countries. Guided by the principles of precaution, differentiated responsibilities, and efficiency, the Climate Convention called for Annex I countries to "take the lead" in controlling GHG emissions and deferred emission limits on developing countries, in recognition of their right to sustainable growth. Annex I differentiation was more fully achieved at Kyoto in 1997, through the ad hoc method described earlier. Many developing countries feared that this would set a precedent that would prejudice any equitable basis for future allocations.

The idea of equal per capita entitlements, however, was refined and developed along the peripheries of the mainstream negotiation process. Various formats, outlined above, evolved over time. The inevitability of developing countries coming into the emissions control regime at some stage and the need for an acceptable strategy for global participation fueled support for and further development of this concept. This includes support within governments. The Indian government was one of the first to officially adopt the equal per capita entitlements approach; at the First Conference of the Parties to the Climate Convention (COP 1) in 1995, the Indian government called for "implementing a program for convergence and sustainable par values for consumption on a per capita basis" (GCI 1999). This was followed by the Africa Group, which presented the Contraction and Convergence proposal in 1997, calling for "reducing the emissions of Annex I" and ensuring a "controlled growth of future emissions of non-Annex I" while being guided by the overall principle objective of "per capita emission rights" (GCI 1999).

The issue of per capita allocations arose repeatedly during the Kyoto negotiations in the context of emissions trading. Some developing countries argued that the emission entitlement or permit should be clearly defined for an initial global allocation before launching any sort of trading. Quite clearly, the system of emissions trading whereby developing countries participate through the Clean Development Mechanism does not define emission entitlements or establish emission rights. China and India called for "equitable allocations" of emission entitlements on a per capita basis as a prerequisite for allowing trading to commence. Subsequently, the European Parliament has also adopted a resolution on climate change, which advocates a global limit of 550 parts per million by volume (ppmv) of CO_2 equivalent, supported by "progressive convergence towards an equitable distribution of emission rights on a per capita basis by an agreed date in the next century." The Indian government has recently reiterated its support for this concept by stating that "equal per capita is an equitable norm and the per capita criterion is central to the determination of emission entitlements" (UNFCCC 2000b). Thus, the concept has managed to progressively expand its support basc in the years since its introduction.

III. Analysis of the Approach: Searching for the Elusive Solution?

Having outlined the evolution and conceptual basis of the per capita entitlements approach, we endeavor in this section to carry out a dispassionate analysis gauging its future applicability within the climate change arena. This is done by organizing a qualitative framework based on some of the key questions surrounding the perplexing issue and analytically addressing them to draw out some useful conclusions.

Is the Atmosphere an Allocatable Natural Resource?

This question forms the conceptual foundation for advocacy of a per capita entitlements approach as well as the basis for establishing and shaping any such entitlement. Thus, the issue merits examination from the outset. Before proceeding to address the issue it is important to clarify the resource in question. Usually termed as the "global atmospheric resource," what it actually alludes to is the "limited assimilative capacity" of the Earth's atmosphere with respect to GHGs.[9] As mentioned earlier, the IPCC provides guidelines on defining the tolerable limits that seek to define this resource, or the GHG assimilative capacity of the atmosphere.

The next paragraphs endeavor to assess whether the "global atmospheric resource" in question can also stand up to some defining attributes. A resource is literally defined as "a stock or reserve, which can be drawn on when necessary."[10] By inference, a resource should possess a reserve "value" and could be quantifiably "drawn on" in case of need (or should possess the capacity of being quantified and allocated).

The first defining attribute, the reserve "value" possessed by any resource, is also linked to the economic concept of scarcity. In the particular case of the atmospheric resource, this scarcity is driven by the idea of a limited assimilative capacity of the atmosphere for anthropogenic emissions of GHGs that carves out this finite resource. As mentioned above, when the atmosphere is discussed in the context of climate change, what really is being discussed is the capacity of the atmosphere to absorb GHG emissions above pre-industrial levels. Also, through international negotiations and the use of climate science and climate models, it can be agreed what level of GHG emissions and associated impacts we as a global community are willing to limit ourselves to and accept. This limitation extends the "scarcity" reserve value to the atmosphere.

Moreover, this scarcity value has been intrinsically recognized, quantified, and capitalized through the scheme of emissions trading. This scheme enables the "atmosphere" to satisfy the second defining attribute, that is, the capacity of being quantified and allocated.

By definition, allowance trading[11] establishes de facto user rights that provide an incentive to protect the environment (Rose and Stevens 1998). Any option to trade is contingent on having first attained this right. Within the climate context, countries are granted the right to consume a certain portion of the limited assimilative capacity of the atmosphere for a certain period of time. The price of emission permits can, thus, be considered as a fee to be paid for the temporary right to use the atmospheric commons beyond its sink capacity. Thus, the price is paid not for owning a piece of the atmosphere in perpetuity, but for obtaining a user right (Ott and Sachs 2000) for a certain predetermined period of time. This user right gives the requisite quantifiable value to the atmospheric resource in question, and implies an ability to support future economic development.

The rights to emit, established by agreements such as the Kyoto Protocol, constitute the practical manifestation of allocating the scarce global atmospheric resource in question and allow it to be quantified for storing, placing in reserve, or banking for future use or sale as and when required. The definitional conditions of a resource thus seem to be met for the global atmosphere. Thus, as outlined above, the global atmosphere can be termed as a resource entailing a "scarcity" value that can be quantified for allocation as well as monetarily capitalized, by utilizing certain economic tools, such as emissions trading.

Is the Equal Per Capita Entitlements Approach Equitable?

Equity may be defined as the "quality of being impartial" or "something that is fair and just" (Banuri et al. 1996). Operationalizing impartiality or fairness in the context of differentiating future GHG limitation commitments has entailed the application of various traditional equity principles.[12] To this end, different overlapping typologies have been used to present relevant equity principles as they apply to questions of distributive justice (IPCC 2001c).[13] As is generally acknowledged, the whole issue of equity cannot be equated only with the principle of egalitarianism, denoting that every human has an equal right to use the atmosphere. Nevertheless, egalitarianism appears consistently in all the research representations, thus establishing it as an important criterion for assessing equity. For instance, Rose and Stevens (1998) include it among five alternate fairness criteria within global warming policy (Table 8.3).

Egalitarianism resonates. It appeals to the hearts and minds of many people the world over. This is why equal per capita entitlements is so consistently recognized. The ethical underpinnings of egalitarian justice are anchored in the international community's ethical standards and legal codes. Prime examples are the United States Constitution as well as the United Nations Charter and its Universal Declaration of Human Rights. All of these furnish strength to the need and desire for applying the egalitarian principle when dealing with the atmospheric common resource, which lies outside the legal purview of individuals or states (Baer et al. 2000):

- The UN Convention on the Law of the Sea requires *common* ownership of deep-sea resources for the benefit of *all humanity* (Articles 16 and 17).[14] This establishes the principle of *joint sharing* of these resources and their associated benefits across the global population.
- Protocol on Environmental Protection to the Antarctic Treaty (Article 7) prevents appropriation of a region's mineral wealth by *any individual nation*. This requires that the common rights to the resource should not be usurped by any individual nation.

Although not explicitly extending equal per capita allocations, the above illustrate how strongly the principle of shared rights and responsibilities applies with respect to managing common resources beyond the territorial jurisdiction of any single country. They not only set the foundations for the advocacy of per capita schemes, but make it difficult to ethically justify any unequal or disparate claims to a global commons, such as the atmosphere.

Table 8.3. A Selection of Alternate Fairness Criteria for Global Warming Policy

Fairness Criterion	Basic Definition	Operational Rule
Sovereignty	All nations have equal rights to pollute and to be protected from pollution	Cut back emissions in a proportional manner across all nations
Egalitarianism	All people have an equal right to pollute or to be protected from pollution	Allow emissions in proportion to population
Ability to Pay	Abatement costs should vary directly with national economic well being	Equalize abatement costs across nations
No Harm	Some nations should not incur costs	Poor countries should not be required to abate emissions
Ad Hoc	Abatement costs should be sensitive to unique circumstances	Give special consideration to economic, health, fossil fuel dependence, etc.

Source: Rose and Stevens (1998).

Overall, equity and fairness principles are indispensable for the establishment and effective sustenance of any global climate change regime. But their practical manifestation has remained uniformly elusive. The absence of a universally established equity doctrine—coupled with the varying economic implications of applying alternate equity principles across countries—has confounded negotiators and advocates with regard to what really is just and fair. As mentioned above, the egalitarian principle does remain a key equity determinant, but at the same time it cannot be the *only* determinant. What the many insightful studies demonstrate is that equity cannot be reduced to any single factor: it is rather a complex concoction of sometimes incompatible, but selectively justifiable, principles.

In this regard, the principle of per capita entitlements possesses a strong and easily comprehensible ethical argument, which lends support to its application. Another important conclusion is that the whole issue of judging equity cannot be selectively narrowed down to the egalitarian principle; it remains just one of the many determinants that can be used for assessing equity within the climate change policy process. Finally, although some valid concerns exist regarding the application of the per capita approach, it remains very difficult to *ethically* justify any *unequal claims* to a global commons such as the atmosphere.

What Are the Linkages with International Emissions Trading?

It is an established fact that the overall cost-effectiveness of international emissions trading is enhanced by increasing the number of participants, and particularly by including those countries (and their private-sector entities) with relatively lower abatement costs. In the context of climate change, this implies that cost-effectiveness is significantly enhanced by, first, using the low-cost mitigation options available in developing countries and, second, committing these countries to emission reduction targets at some future date (Rose and Stevens 1998). In this context, an allocation rule that catalyzes the above as early as possible would improve the cost-effectiveness that emissions trading brings to participants.

In terms of relative benefits, many developing countries generally stand to gain more, as already noted, under a per capita allocation regime than under other approaches. It follows, logically, that they may be much more inclined to join such a regime. Thus, it is likely that a per capita scheme would be able to attract the earliest possible entry of developing countries, thereby allowing emissions trading to maximize net benefits. This has been corroborated by various models, which suggest that per capita (convergence) offers the best opportunities for capitalizing on cost-reduction options as all Parties can fully participate in emissions trading (Berk and Elzen 2001).

Another pertinent observation emerging from various analyses is that trading is not merely *good* for the per capita scheme, but also that an equal per capita entitlements approach is *no good* without trading. A recent modeling study (Bohringer and Weisch 2000) indicates that the per capita approach entails significant global welfare costs that, without trading, can be several times higher than other allocation schemes. Chapter 9 finds similar results when focusing on emission reduction costs instead of welfare costs. Thus, trading has a direct bearing on the efficacy and efficiency of the per capita approach.

Thus, whereas the per capita approach has the potential to maximize the trading benefits through early developing-country participation, the inclusion of trading is also deemed essential for enhancing the relative success and appeal of this approach. Given this finding, the chapter now addresses some other associated questions.

Concerns about hot air? Within any trading regime, the potential exists for issuance of excess emission allowances above business-as-usual levels to certain countries (popularly termed "hot air"). Such inflated targets could threaten to undermine the environmental integrity of the system and allow for reductions on paper rather than actual carbon-reducing trades. However, as stated, this issue is associated with allocation and trading in general, and is not specific to the per capita approach. In fact, the potential for creating hot air in a negotiated regime like the Kyoto Protocol or an approach seeking the voluntary opt-in of developing countries (Aslam 2001) is even more problematic than under an approach like equal per capita entitlements. A number of factors are at play.

First, the per capita approach is an objective one that allocates entitlements according to a formula based on two indicators: population and emissions. Both these indicators are widely used and reported, and there is a limited scope for any major manipulation. Second, the chosen target in the case of a per capita approach is one of annual contraction as well as cumulative convergence that is similar for all the participating entities. Third, hot air is largely viewed as a political incentive that could extend relaxed and generous commitments to developing countries in exchange for early participation. While hot air can be used to provide political incentives for participation within a negotiated or voluntary participation regime, it is at least controlled and limited in an equal per capita entitlements approach seeking simultaneous participation from large blocs of developing countries. Thus, the chance of "negotiated" manipulation is significantly reduced when dealing with a top-down approach like per capita that apportions entitlements based on an indicator and strives for global participation.

Finally, corrective measures can still be undertaken to better manage hot air, such as giving the suspect countries the right to trade away only a certain fixed percentage of their unused emissions while banking the rest for future domestic use (Agarwal et al. 1999).

Associated obligations? The inclusion of developing countries in an emissions trading regime would not come without a cost. Trading is a conditional and not an absolute right. Instituting and implementing a domestic system to conform to the dictates of the international trading regime would carry a number of associated requirements and obligations, such as emission measurement and verification. The Kyoto Protocol stipulates most of these requirements, which Box 8.1 outlines.

What is certain is that significant domestic human and institutional capacity would be required in order to conform to these emissions trading requirements. Also, this capacity would need to be supported with available, reliable, and credible emissions data, capable of withstanding international audit and scrutiny. All of this poses a challenging task, even for

Box 8.1. Treaty Obligations for Countries Engaging in International Emissions Trading

Greenhouse gas emissions target. The allocation of the right to emit automatically creates an obligation not to exceed that limitation. So, all participating countries would have an emission limit set over specified time frames, and this obligation must include some mandatory consequences for noncompliance.

National systems for greenhouse gas inventories. Such systems for inventories need to stipulate institutional arrangements, quality control mechanisms, information management systems, reporting systems, etc. The system must also require the country to accommodate audits and inspections.

National registries would be required to record and track the transfers and acquisition of emission allowances.

Reporting. Countries would need to submit annual greenhouse gas inventories according to agreed international standards. In addition, countries need to supply other important information, such as descriptions of their national inventory systems and national registries as well as transactions undertaken through international emissions trading or project-based mechanisms.

Review of information. Countries would need to provide auditing teams with data and information necessary for assessing whether the country has conformed to technical requirements and international standards of good practice in their national systems, reporting, and registries.

***Additional domestic regulations.** To the extent that countries wanted to allow domestic companies to participate in international emissions trading, additional domestic laws or regulations would be required. Participating companies would be subject to emission limitations and corresponding responsibilities to measure and report emissions in a standardized manner.

Source: Baumert et al. (2002).

developed countries. This issue is not directly related to the per capita approach but, as stated, becomes relevant owing to the fact that trading is essential to ensure the effectiveness and successful implementation of the per capita approach.

Some of the obligations, however, such as monitoring and reporting of emission inventories under the National Communications, are already obligations for developing countries under the Climate Convention, although with a lower degree of intensity and enforcement. The advent of emissions trading would naturally raise the requirements for ensuring credibility and transparency.

Finally, emissions trading is firmly established as the instrument of choice within the current climate mitigation architecture, as outlined under the Kyoto Protocol. As such, its application within any other future expansion regime can also be logically expected. Thus, prior to burdening any trading-related liability on the per capita approach, it would be essential to investigate the consequences of avoiding trading under the other possible allocation approaches as well as assessing the comparative data and institutional requirements for them. This has to be an essential consideration for any comparative analysis of the allocation schemes.

In any case, owing to the nascent state of the carbon market and the requirement to build it from scratch, the costs associated with instituting the requisite architecture could constitute a significant barrier in the short to medium term. The capacity to respond to these obligations would depend on a host of factors influenced mainly by national circumstances, such as political will, domestic preparedness, and the possibility, certainty, and extent of any financial benefits. As the analysis has already suggested, there are significant country variations with respect to per capita emissions. Thus, a scheme of temporal graduation could enhance the acceptability and institutionalization of these obligations in the long term.

Overall, this section's analysis testifies first that a successful and effective implementation of per capita-based regime is linked inescapably to international emissions trading, which is deemed essential to ensure cost-effectiveness and environmental efficiency. Secondly, there exist certain issues of concern related to trading per se, which are not solely a challenge for this particular approach. However, given the strong reliance upon trading, any risk of trading failure could be accentuated under this scheme, to the detriment of the environmental effectiveness and cost-effectiveness of the regime.

Is the Approach Flexible?

Owing to the nature of the climate change issue and its associated complexities and uncertainties, any effective approach needs to be flexible to both incorporate any future scientific developments as well as accommodate the disparities among countries. As already discussed, a per capita convergence approach is able to readjust to tighten or relax yearly contraction budgets, as well as realign its overall reduction trajectory (convergence) to respond to any change in carbon concentration target.

However, the per capita approach's flexibility to account for the differing national circumstances is limited. Quantified emission levels, such as the ones used for the per capita approach, do not account for the social quality of these emissions—that is, to distinguish between "luxury" and "survival" emissions (Agarwal et al. 1999). Similarly, such simplified indicators also fail to consider factors such as geographical/climatic conditions or the structure of the respective economy and energy supply, each of which has an important bearing on the variance of emissions among countries. For instance, a high endowment of hydro resources (e.g., in Norway and Brazil), high dependence on nuclear energy (e.g., in France), a high level of industrial efficiency (e.g., in Japan), or an exceedingly cold climate (e.g., in Iceland) can have correspondingly favorable or adverse influences on the per capita emission levels. The per capita approach does not address these disparities, and potentially creates unwanted distortions, such as taxing countries with efficient economies or punishing countries with limited access to renewable resources (e.g., hydropower) that would tend to reduce their emission levels.

In the presence of large differences between countries, this limitation can be a major factor impeding progressive acceptance of equal per capita entitlements. Recognizing this shortcoming, some proposals deviate from the pure per capita convergence approach (Gupta and Bhandari 1999).[15] These proposals adjust the approach through the inclusion of allowance factors, such as those mentioned above, which can allow for country-specific characteristics that contribute significantly to variations in emissions per capita (Ybema et al. 2000). The concept of graduation, which allows increasing participation in this regime, does offer a restrained enhancement of flexibility in terms of different starting points.

Is the Approach Consistent with the Provisions of the Climate Convention and Kyoto Protocol?

As outlined earlier, the Climate Convention and the Kyoto Protocol provide the policy framework for international cooperation in the field of climate change. Any differentiation proposal, such as the per capita approach, needs to be consonant with the Convention's basic provisions, including the ultimate objective of trying to "stabilize greenhouse gas concentrations in the atmosphere at a level that would prevent dangerous anthropogenic interference with the climate system." In addition, other pertinent principles of the Convention guiding any expansion of commit-

Box 8.2. Guiding Principles Established Under the Climate Convention

- All Parties to act "on the basis of equity and in accordance with their common but differentiated responsibilities and respective capabilities."
- Give full consideration to Parties that "would have to bear a disproportionate or abnormal burden under the Convention."
- Developed-country Parties to take the lead.
- Right to promote sustainable development.
- Allow for growth of the share of global emissions from developing countries.
- Strive for the widest possible cooperation.
- Account for specific needs and special circumstances of developing countries and vulnerable parties.
- Developing-country Party commitments conditional upon successful developed-country implementation of commitments related to financial resources and technology transfer.

Source: UNFCCC (1992, Articles 3 and 4).

ments as well as developing-country participation are summarized in Box 8.2.

The per capita approach begins by setting an ultimate collective target, based on current IPCC estimates. Thus, it endeavors to conform to the ultimate objective of the Convention while retaining a readjustment flexibility, which aims to ensure that the environmental target is ultimately met.

The approach then applies the principles of "common but differentiated responsibilities" and "need for growth of developing country emissions" by allowing most developing countries' per capita emissions to grow, while demanding a reduction of most developed countries' per capita emissions. Thus, it also encourages the developed countries to "take the lead" in cutting their emissions while extending space for sustainable growth of the developing countries.

The approach in its pure form, however, falls short when it comes to accounting for particular country circumstances, such as difference in Parties' "starting points and approaches, economic structures and resource bases, the need to maintain strong and sustainable economic growth, available technologies and other individual circumstances" (UNFCCC 1992,

Article 4.2(a)). As stated earlier, some variants of the approach have endeavored to address this shortcoming by phasing the participation of both developed and developing countries. However, none of these variants has, so far, managed to successfully work across the diversity of country circumstances and enhance the political palatability of this approach.

Thus, as far as the Convention is concerned, the approach adheres to most of the guiding principles but falls short when it comes to incorporation and consideration of particular country circumstances. Along with the Convention, however, it is also important to assess whether it can carry the Kyoto baggage described earlier.

If the Protocol enters into force, the precedent of ad hoc quantitative targets based on politically negotiated justice and the absence of any objective formula is the baggage that would need to be accommodated by any expanding regime. The Kyoto architecture, along with the associated use of international emissions trading, should be practically acceptable, at least in the short to medium term. Thus, any effective regime for expanding participation and commitments should be able to amalgamate this reality.

The per capita approach is not at odds with such an eventuality. In fact, it offers the possibility of a two-track approach.[16] Some have proposed continuation along the Kyoto track for Annex I until after the first or second commitment periods, while allocating per capita-based entitlements to the developing countries. Alongside this, the Kyoto-based allowances, already apportioned to Annex I countries, and the per capita entitlements could be fungible to promote cost-effectiveness through emissions trading. Such a two-track approach allows for a "soft transition" of Annex I countries while also allowing for the possibility of a "phased graduation" for developing countries. Various other adjustments (such as use of an efficiency index) have also been suggested and were elaborated on earlier. Also, as described above, the market-based architecture enshrined in the Kyoto Protocol through the concept of emissions trading can be adopted by an equal per capita entitlements approach. In fact, doing so is essential to the successful implementation of this approach. Thus, it promotes and reinforces the market-based framework, which forms the linchpin of the Protocol.

The per capita approach, thus, has the design capacity to carry the Kyoto baggage and does not necessarily demand a revolutionary revamping of the current architecture, but rather a gradual amalgamation toward eventual equal per capita entitlements.

What Is the Potential for Global Acceptability of the Approach?

Broad global acceptability is logically considered a prerequisite to the success of any approach for differentiation of future commitments. The potential for acceptability is, however, determined by a combination of factors having political and ethical as well as economic dimensions. Egalitarian equity, as shown above, formulates the ethical basis for this judgment. The other factors are mostly influenced and driven by national interest and circumstance, which may not always be aligned with accepted ethical norms. Negotiation theory, in fact, generally assumes that actor behavior is primarily motivated by self-interest, and suited principles of fairness are selectively invoked in order to defend this interest. Thus, to gauge the potential for global acceptability of this approach, it would be useful to explore the nexus of self-interest and fairness. Doing so can help judge both the diversity among states as well as its influence on their views about equity within climate negotiations.

It is no secret that countries differ greatly in terms of size, resource endowments, population, wealth, GHG emissions, vulnerability, and ability to respond to climate change. Table 8.4 shows some of the disparities between countries, including emissions per capita, which are generally much higher in the developed countries than in the developing ones.

Given these disparities, various studies and models have analyzed the impacts of different burden-sharing rules on a country or regional basis.[17] Although various models use different time frames, parameters, and methodologies, they can nevertheless provide some general indicators of the outcomes. The comparative analyses suggest that benefits are likely to be skewed within the per capita approach. Appendix 8A provides comparative results of three such studies with regard to their application to the per capita approach. Countries with large and growing populations or with low emissions stand to benefit more than the others. In all cases, however, China, India, and sub-Saharan Africa (excluding South Africa) come out as consistent gainers under a per capita approach. The oil-producing and more developed of the developing countries—such as Singapore, United Arab Emirates, Argentina, and South Africa—are relatively disadvantaged among the non-Annex I countries. Chapter 9 shows similar results. It comes as no surprise that some of these developing countries are not vociferous advocates of this approach.

As discussed above, the major proponents of the per capita approach in climate change negotiations have been India, China, and Africa, which also happen to be the major beneficiaries. With less intensity and greater ambivalence, the European Union and France have argued for a long-

Table 8.4. Regional Variation in Key Economic and Emissions Indicators, 2000

	Population (millions)	Income Per Capita (PPP, int'l dollars)	Carbon Emissions Per Capita (tons)
World	**6,057**	**7,415**	**1.1**
Annex I	**1,170**	**22,377**	**3.3**
United States	286	33,633	5.6
European Union	378	23,612	2.4
Japan	127	26,755	2.5
Eastern Europe	281	7,926	2.5
Non-Annex I	**4,888**	**3,834**	**0.5**
China	1,269	4,089	0.6
India	1,016	2,358	0.3
Other Asia	1,103	6,225	0.8
South Africa	43	9,401	2.5
Sub-Saharan Africa	659	1,598	0.2
Argentina	37	12,377	1.0
Latin America	512	7,181	0.7

Source: Based on data from EIA (2002b) and World Bank (2002).
Notes: Includes carbon emissions from fossil fuel combustion only.
Abbreviation: Purchasing Power Parity (PPP).

term convergence toward equal per capita entitlements, while Japan has advocated it as one of the two indicator options to choose from, in its proposals at various COP meetings (see Table 8.1). While not beneficiaries vis-à-vis India, China, and Africa, these countries do come out as a relative beneficiaries within the context of Annex I due to their relatively low per capita emissions (about half the per capita emissions of the United States and lower than the Annex I average; see Table 8.4). This makes it relatively convenient for these countries to embrace an otherwise strong ethical position. The main opponents of the scheme would likely be the United States and Russia, as they would carry the brunt of the wealth transfer from any shift to equal per capita entitlements (Table 8.5).

Table 8.5 suggests the presence of a positive nexus between national self-interest and the choice of equity principle. This analysis suggests that each country can be expected to argue for a scheme that suits its national circumstances, which effectively selects the preferred equity. In addition, owing to the fact that some relative beneficiaries suggest this approach, there is evidence to suggest an acceptable ethical foundation of equal per

Table 8.5. Self-Interest ~ Fairness Nexus

Country / Region	Advocate of Per Capita approach	Beneficiaries
China	Yes	Yes
India	Yes	Yes
Africa	Yes	Yes
Japan (choice option)	Yes	Relative Yes
France (long-term convergence)	Yes	Relative Yes
European Union (long-term convergence)	Yes	Relative Yes
United States	No	No
Former Soviet Union	No	No
South Africa	No	Relative No
Argentina	No	Relative No

Notes: The positions indicated above are based on the Kyoto Protocol negotiation process (Ad-Hoc Group on the Berlin Mandate).

capita entitlements, especially as a long-term guiding principle. In the short term, however, the significant disparities between countries challenges the notion of applying simple, fixed, top-down allocation schemes, such as the per capita allocation. The analysis highlights the difficulty of proposing a single composite formula that can satisfy the strong and diverse national self-interests.

Thus, in the short to medium term,[18] it seems to be politically unrealistic and procedurally difficult to adopt a rule such as equal per capita entitlements. Even with a strong ethical foundation, it runs counter to the self-interest of some pivotal actors, such as the United States, Russia, and parts of the OECD. The potential of its acceptability by a critical mass of actors within the climate negotiations process is, therefore, likely to be limited in the short to medium term. This is especially the case because a few key participants can hold the process hostage. With some key negotiating countries, such as the United States, Australia, and Russia, not likely to be the main beneficiaries under this approach, as earlier indicated, the potential for stalling the process remains threateningly present. This threat becomes all the more potent owing to the *procedural* rule in Climate Convention decision-making, which is based on consensus (i.e., the absence of dissent).

Would Equal Per Capita Entitlements Encourage Population Growth?

Before closing, it is worth addressing the concern expressed by some critics that equal per capita entitlements would promote population growth. This concern stems from the fact that per capita entitlements, by compensating for large populations, may deliver more entitlements to countries with increasing populations. (To alleviate this concern, proponents advocate the use of a population base year.)

At a fundamental level, this concern seems to be based on flawed assumptions. First, the notion that additional "entitlement dollars" would offset all the other economic repercussions of burgeoning populations is far-fetched. Any effect by allocation of entitlements could be negligible compared to other factors necessitating population control, such as poverty alleviation and resource constraints. Second, any enhancement of the entitlement quota would imply a correspondingly larger number of people sharing it (Agarwal et al. 1999), which could quickly outstrip any associated benefits. Of all the concerns about equal per capita entitlements, this seems the least significant.

IV. Conclusions

As the above analysis shows, the per capita approach endeavors to bring a multidimensional solution to a complex problem. The merits and demerits of the equal per capita entitlements approach are summarized below.

Merits

- Simplicity of concept
- Strong ethical basis
- Flexibility to accommodate changing scientific evidence
- Enhancement of efficiency of global trading
- Offer of incentives for developing-country participation
- Consistency with the major guiding principles of the UNFCCC
- Amalgamates well with the Kyoto architecture

Demerits

- Limited global acceptability
- Limited flexibility for accommodating varying country circumstances
- Linkage with trading essential for success
- Associated issues of hot air and obligation costs

Given the above demerits, a number of variants to the pure per capita approach have been suggested. The primary aims of such modifications have been to enhance global acceptability and accommodate varying country circumstances by extending a transition phase or phasing out requisite emission reductions through some "allowance factors" or "soft landing scenarios" (Blanchard et al. 2001). So far, however, none of these adjustment proposals has been able to delicately balance the conflicting interests and ethical philosophies in a politically palatable fashion.

Nevertheless, per capita's ethical foundation has a strong defining power, which is likely to shape long-term approaches, and we can certainly expect the concept to be invoked with a growing degree of legitimacy in the future. However, its time may not yet have arrived. In the short to medium term, the process may need to be more adaptive rather than immediate (Toman and Cazorla 2000); during this time frame, political realism tends to lead toward a system of "adjusted egalitarianism" (Ott and Sachs 2000). Such a scheme could include explicit provisions for country-specific circumstances, as well as possible amalgamation with some of the other proposals for expansion of commitments, in an effort to enhance global appeal. In the long term, the idea has the potential to be a *guiding principle* toward an eventual convergence of global per capita emissions—if not on an absolutely equal level, then at least within the confines of a defined and globally accepted "sustainability corridor" (TERI 1997).

Given the constraints cited above, it might be possible to constitute a future GHG emissions entitlement by combining some of the merits of the per capita approach with other approaches in an endeavor to overcome some of the stated shortcomings. In this respect, a practical manifestation of an entitlement of the future could be envisioned, as shown in Box 8.3.

Such a compromise could begin by first defining and then quantifying a per capita level of

Box 8.3.
Emission Entitlement
of the Future?

VARIABLE
portion accounts for country-specific circumstances and other approaches

FIXED
per capita entitlement portion based on allocation of "survival" emissions

survival emissions that are required by each human being to sustain a reasonable standard of living. This minimum per capita level could then be distributed across the countries according to their respective populations and would provide the fixed portion of the entitlement. The remaining flexible, varying portion of the entitlement could then be defined by accounting for particular country circumstances.[19] This accounting and quantification would also need to utilize other useful approaches in order to gain maximum political acceptance across the globe. The above are just some preliminary thoughts on what could possibly shape a future emission entitlement—driven by environmental effectiveness, motivated by economic efficiency, and packaged by political compromise.

Appendix 8A.

Table 8A. Comparative Analysis of Per Capita Allocations

Country/ Region	Egalitarian Allocation-1 (2020-Bn of US$1990)[1] *estimated minimum cost to implement per capita allocation of carbon emissions in 2020*	Egalitarian Allocation-2 (C/pop in %)[2] *estimated percentage of global per capita allocation of carbon emissions*	Egalitarian Allocation-3 (% of 1995 emissions)[3] *estimated per capita allocation of carbon emissions in 2015*
United States	354.5	4	
Japan		2	
Canada/Western Europe	29.9		
European Union		5	
Other OECD	65.3	2	
Eastern Europe/ Former Soviet Union	345.5	2	
China	–109.1	20	145
Middle East	1.1		
Energy-exporting countries		16	
Africa	–226.3		
Nigeria			466
South Africa			51
Latin America	56.6		
Brazil		3	237
Argentina			102
Southeast Asia	37.2		
Dynamic Asia		3	
India	-----	16	382
Rest of World		22	

Sources: [1]Rose et al. 1998; [2]Reiner and Jacoby (1997); [3]Winkler et al. (2001).
Abbreviation: Organization for Economic Cooperation and Development (OECD).

Egalitarian allocation-1 enlists the results of a model that attempts to quantify the minimum cost of implementing the per capita allocation rule. The analyses of the cost allocations were performed for three future years (2005, 2020, and 2035) and the costs were discounted to a 1990 present value. As the table shows, the model indicates widely disparate outcomes, with wealth transfers mainly from the United States and the former Soviet Union to Africa, China, and Asia.

Egalitarian allocation-2 starts by totally discounting historical emissions and then allocating permits according to the per capita approach. However, even after discounting all historical liability, the model suggests that the United States and OECD countries would receive a small percentage of the global allocation, with the major share going to India and China.

In egalitarian allocation-3, the model compares the outcome of the allocation based on equal per capita across five developing countries in the year 2015 for a global reduction target of 4 percent below 1995 levels. The results are shown as a percentage of 1995 emissions. The outcome suggests a wide variance even among developing countries, with the more developed but less populated countries—South Africa and Argentina—receiving relatively fewer entitlements. (South Africa is, in fact, in deficit compared to its 1995 emissions level.) On the contrary, the more populated countries receive higher entitlements.

Notes

1. However, the "common but differentiated" and "polluter pays" principles formed a solid basis for differentiation between Annex I and non-Annex I countries.
2. Ad Hoc Group on the Berlin Mandate.
3. To stabilize GHG concentrations at a level that would prevent dangerous anthropogenic interference with the climate system.
4. Such as the "sinks" scheme, advocated by the Center for Science and Environment, which distributes the estimated global GHG absorptive capacity, as well as the "moving" entitlements scheme, which assigns ad hoc an initial per capita entitlement subjected to periodic reviews.
5. Second World Climate Conference.
6. Such as France, Switzerland, and the European Union. See Torvanger and Godal (1999).
7. One the most widely quoted scenarios sets the level at a concentration target of 450 ppmv (parts per million by volume) of CO_2 by 2100.
8. Although both advocate the idea of convergence, there is a difference in the approaches. While GCI counts convergence to an equal level of emissions as the final goal of a per capita framework, CSE views the approach as a means to incentivize a transition to renewable energy technologies in developing and industrialized countries.
9. Thus, the term "global atmospheric resource," as used in this chapter, always implies the GHG assimilative capacity of the atmosphere.
10. Müller (2001a). The source of definition is the *New Shorter Oxford English Dictionary*.
11. The kind of trading allowed between Annex 1 countries within the Kyoto regime and what is suggested as the associated trading regime with "per capita" entitlements scheme.

12. In the climate change arena, the issue of equity has been focused primarily within the narrow context of the differentiation of future commitments. This focus is inequitable itself, as it unjustly ignores the issue of "adaptation burdens," which is crucial for a large number of vulnerable developing countries. However, in the context of this analysis, this chapter will focus on the application of equity to the expansion of future developing-country participation and allocation of mitigation burdens as they relate to the per capita approach.
13. This refers to the allocation of collective benefits and burdens among the members of a community on local, national, or global levels. See IPCC (2001c) for a comprehensive listing of the various equity typologies.
14. "Desiring by this Convention to develop the principles embodied in resolution 2749(XXV) of 17 December 1970 in which the General Assembly of the UN solemnly declared inter alia that the area of the sea-bed and ocean-floor and the subsoil thereof, beyond the limits of national jurisdiction, as well as its resources, are the common heritage of mankind, the exploration and exploitation of which shall be carried out for the benefit of mankind as a whole, irrespective of the geographical location of states."
15. The Tata Energy Research Institute (TERI) scheme mixes per capita allocated entitlements to non-Annex I countries with reductions to Annex I countries adjusted by an efficiency index.
16. For instance, TERI, CSE, and GCI all propose this sort of a "two-track" approach.
17. Toman and Cazorla (2000) provide a summary of the various analyses.
18. "Short to medium term" can be taken to mean until the second commitment period.
19. UNFCCC Article 4.2(a) could be used as guidance.

9. Scenarios for Differentiating Commitments:

A Quantitative Analysis

Odile Blanchard

As emphasized in the latest Intergovernmental Panel on Climate Change (IPCC) report, stabilization of greenhouse gas (GHG) atmospheric concentrations is needed in order to delay and reduce damages from climate change (IPCC 2001d, Q6.9). The previous chapters in this volume qualitatively discuss paths toward future global participation in this effort to mitigate climate change. Various proposals are analyzed, ranging from principle-based allocation methods to more pledge-based, country-tailored approaches. All of them could contribute to achieving the ultimate goal of the 1992 United Nations Framework Convention on Climate Change (UNFCCC), which is to stabilize GHG concentrations at a safe level.

This chapter examines three worldwide scenarios of differentiated commitments from a quantitative perspective. Each scenario is drawn from a proposal analyzed in this volume. The Per Capita Convergence scenario allocates emission allowances to countries based on population. The Relative Responsibility scenario shares emission reductions according to the countries' respective responsibilities for climate change. The Emissions-Intensity Target scenario frames the mitigation effort on the basis of reductions in carbon intensity. Based on a long-term concentration stabilization goal, each scenario focuses on the period 2010 to 2030. The POLES model (described in Appendix 9A) is used to investigate the carbon dioxide (CO_2) emission limitations needed to meet an intermediary environmental goal in 2030, and their distribution across countries.[1]

This chapter shows how differentiating commitments based on various proposals could be translated into operational terms and used to induce Annex I countries to further take the lead in a global participation framework to limit CO_2 emissions. It illustrates several issues raised in the previous chapters and may provide useful information for future climate change negotiations. The findings show how the three differentiation scenarios yield varying CO_2 emission allowances and abatement costs across countries.

Section I presents the assumptions and methodology used in the analysis throughout the chapter. Sections II, III, and IV discuss the distribution of emission allowances implied by each scenario. Section V compares the scenarios with respect to emission allowances and permit trading.

I. Assumptions and Methodology

The environmental goal at the center of this analysis is stabilization of atmospheric CO_2 concentration between 450 and 550 parts per million by volume (ppmv) by 2100. This long-term target range corresponds to the lowest CO_2 concentration targets adopted in the emission mitigation scenarios examined in the latest IPCC report (IPCC 2001c). This represents, at most, a *doubling* of CO_2 atmospheric concentrations compared to pre-industrial levels. Stabilization at such a level could still entail potentially serious damages attributable to climatic changes.

Consistent with a 450 to 550 ppmv CO_2 stabilization goal, an intermediary goal is set at 9.4 billion tons of carbon equivalent (GtC) for 2030. This constitutes the *maximum* level of annual world CO_2 emissions. The reason for this intermediary target stems from the trajectories for CO_2 emissions from fossil fuel combustion drawn in the IPCC Third Assessment Report, from 1990 onward (IPCC 2001c). To achieve concentration stabilization between 450 and 550 ppmv, most of the fossil fuel CO_2 mitigation scenarios reviewed by the IPCC show similar inverted U-shaped emission trajectories. The inverted U-shaped emission trajectories mean that, after a period of growth, emissions reach a maximum between 2020 and 2060, stabilize for a time, and finally decline at a different rate (IPCC 2001c, 130, 150). The maximum emission level ranges from 6 to 15 GtC per year in the IPCC review. The maximum emission level of 9.4 GtC in 2030 is in the lower range of these trajectories.

This analysis uses the POLES model, a partial equilibrium model of the energy sector, and the ASPEN software.[2] Thus, only CO_2 emissions from fossil fuel combustion are taken into account.[3] Countries are considered either individually or on a regionally aggregated basis. The assumptions for 2010 emissions are designed to reflect current conditions (Box 9.1), with global emissions reaching approximately 7.8 GtC in 2010. Between 2010 and 2030, CO_2 emissions increase from 7.8 GtC to nearly 12 GtC. This reflects a business-as-usual (BAU) trajectory, whereby no action is taken to mitigate CO_2 emissions.[4]

Box 9.1. Assumptions for 2010 Emission Levels

- All *Annex I countries* except the United States and the economies in transition are assumed to reach their Kyoto targets.
- *The United States* is assumed to achieve the Bush administration's target, which is to cut the greenhouse gas intensity of economic production by 18 percent between 2002 and 2012 (White House 2002).
- The emissions of the *former Soviet Union and other Eastern European countries* are assumed to equal the business-as-usual (BAU) projections (which are far below their Kyoto targets, due to economic slowdown).
- *Non-Annex I emissions* follow the model's BAU projections, because non-Annex I countries do not have binding targets under the Kyoto Protocol.

The intermediary environmental goal of limiting global emissions to 9.4 GtC by the year 2030 thus represents an overall reduction of emissions of nearly 2.6 GtC in 2030, relative to the BAU case. This chapter examines how this reduction may be distributed among countries according to the three differentiation scenarios and the associated economic outcomes. As with most models, the results in the following sections are best interpreted in relative terms rather than absolute figures.

II. The Per Capita Convergence Scenario

As pointed out in Chapter 8 of this volume, the distribution of emission allowances based on a per capita rule is a resource-sharing issue, namely, a global emission budget to be equally allocated among all the people of the world. The Global Commons Institute and the Center for Science and Environment have played important roles in developing and advocating per-capita-based approaches since the early 1990s (See Meyer 2000, Agarwal and Narain 1991).

Given the wide discrepancies among countries' current levels of per capita emissions, convergence to an equal per capita allowance level may require a few decades to become politically acceptable to today's high per capita emitters. The Per Capita Convergence scenario sets the emission convergence year at 2050, meaning that by then, per capita emission allowances will be the same in all countries: 0.95 tons of carbon equivalent (tC) per year. The transition period—2011 to 2049—is divided into three

Box 9.2. Countries' Emission Allowances Under the Per Capita Convergence Scenario

For each year, the calculation is completed in two steps:

1) Calculation of country i's emission share
The calculation is based on one of the equations proposed by the Global Commons Institute (GCI 2002) to achieve convergence to a standard value.[1]

$$S^i_y = S^i_{y-1} - (S^i_{y-1} - P^i_y)* \exp^{(-a*(1-t))}$$

where
S^i_y is the emission share of country *i* in year *y*.
P^i_y is the population share of country *i* in year *y*.
a is the "convergence coefficient" (set to 4). The higher the value, the later the convergence occurs. Setting *a* to 4 corresponds to a convergence trend beginning between 2020 and 2030.
t is the elapsed time ratio between starting year (2011, t=0) and convergence year (2050, t=1).

2) Calculation of country i's emission allowance

$$A^i_y = S^i_y * B_y$$

where
A^i_y is the emission allowance of country *i* in year *y*.
S^i_y is the emission share of country *i* in year *y*.
B_y is the global CO_2 emission budget of year *y*.

Source: Adapted from GCI (2002).

[1] The exponential convergence function was chosen in the present Per Capita Convergence scenario because it makes the transition smoother in the early years. GCI also proposes a linear convergence function, which is simpler and "removes the arbitrary and possibly contentious speed-of-convergence parameter 'a' from the model." See GCI (2002).

subperiods and reflects the most common curve identified by the IPCC for emission trajectories in concentration stabilization scenarios. From 2010 to 2030, global emissions grow in a linear fashion from 7.8 to 9.4 GtC. They stabilize for the next 10 years and finally decrease by 1 percent per year from 2040 to 2050. Thereafter, these yearly carbon budgets are then allocated to countries on the basis of population. The calculation of a country's emission allowance is described in Box 9.2.

Table 9.1. Emission Allowances in the Per Capita Convergence Scenario

	BAU	*Per Capita Convergence Scenario*	
Countries	**2030 Emissions (MtC)**	**2030 Allowances (MtC)**	**Reduction (–) or increase (+) in emissions relative to 1990 (%)**
Annex I			
United States	1,951	878	–34
European Union	1,067	598	–31
Japan	331	202	–31
Australia and New Zealand	158	64	–19
Former Soviet Union	944	466	–51
Other Economies in Transition	282	181	–34
Annex I, all others	210	90	–35
Non-Annex I			
Brazil	226	230	+322
Mexico	183	156	+93
India	1,180	1,333	+713
South Asia, excl. India	179	501	+2130
China	2,395	1,777	+173
South Korea	249	96	+48
Southeast Asia	921	789	+270
Africa	716	1,231	+626
Gulf States	473	293	+111
Non-Annex I, all others	516	516	+195
World	11,981	9,400	+66

Source: Calculated using POLES model.
Note: See Appendix 9B for the definition of geographic regions.
Abbreviations: Business as usual (BAU), millions of tons of carbon equivalent (MtC).

Table 9.1 shows the emission allowances by 2030 and compares them with BAU emission levels and the relative change since 1990. As expected, considering their current high emissions per capita, Annex I countries would have to considerably reduce their emissions relative to both the BAU case and 1990 levels.

Within the non-Annex I countries, the situation is different: The 2030 allowances would be greater than the 1990 emissions, which would give them an opportunity to expand their economies and subsequent emissions to meet some of their development expectations. Some of the countries would need to reduce their emissions compared to 2030 BAU levels.[5] Others, currently low per capita emitters, would have allowances greater than the 2030 projected BAU emissions. In other words, they would be allocated more emission allowances than the model projects they will need. This is due to a combination of factors, such as their low current levels of emissions per capita, their future economic prospects as estimated in the model, their population growth pattern, and the level of the convergence target by 2050.

III. The Relative Responsibility Scenario

The Relative Responsibility scenario distributes required yearly emission reductions according to an indicator of relative responsibility for climate change. This scenario is similar to the variant of the Brazilian Proposal suggested in Chapter 7 of this volume; it defines responsibility on the basis of *cumulative* CO_2 *emissions*, rather than in terms of contribution to global warming, as in the original version of the Brazilian Proposal. Here, cumulative emissions since 1900 are used as a proxy to assess historical responsibility for global warming, partly because the CO_2 emissions estimates exist (from fossil fuel use only) and may be used with reasonable confidence.[6] In addition, as pointed out in Chapter 7, "expressing responsibility in terms of *cumulative emissions* over time... reduces the need for complex scientific models and associated uncertainties...." Relative to the Brazilian Proposal approach to determining responsibility, this scenario places a somewhat greater burden on countries that industrialized early.[7]

Several other features of this scenario are also consistent with modifications of the original Brazilian Proposal as suggested in Chapter 7. First, this scenario is applied to all countries of the world, not just industrialized countries. Second, it is based on emission reductions *relative to the BAU case* to reach the 2030 environmental target of 9.4 GtC. This target was chosen because the original Brazilian Proposal is only conducive to absolute emission *reductions* (not to increases), and the 2030 intermediary environmental goal used here leads to an *increase* in emissions relative to 1990.

The yearly global emission budgets are the same as those used for the Per Capita Convergence scenario.[8] This allows a comparative analysis between scenarios. The yearly global CO_2 emission reductions are thus the difference between the global BAU yearly emissions and these yearly global budgets. The yearly global reductions are then distributed to each country in proportion to their relative responsibility for CO_2 emissions since 1900 (see Box 9.3).

Box 9.3. Distributing Emission Reductions Based on Relative Responsibility: An Example

The following steps are used to calculate the emission reductions that the United States would have to achieve in 2030.

1) Calculating estimated 2030 global CO_2 reductions:

2030 business-as-usual (BAU) global emissions (POLES model) = 11,981 million tons of carbon equivalent (MtC)
2030 emission budget = 9,400 MtC
2030 global reductions = 2030 BAU global emissions – 2030 emission budget = 11,981 – 9,400 = 2,581 MtC

2) Calculating 2030 U.S. relative responsibility:

U.S. cumulative emissions from 1900 to 2020: 111.3 GtC
World cumulative emissions from 1900 to 2020: 421 GtC
U.S. cumulative emissions as percentage of the world cumulative emissions from 1900 to 2020 = 26. 43%

3) Calculating the U.S. emission reductions relative to BAU in 2030:

It is the product of global reductions and the ratio of U.S. relative responsibility = 2,581 * 26.43 % = 682 MtC

Relative responsibility is measured as the ratio of the cumulative emissions of a country to the world cumulative emissions. As in the original Brazilian Proposal, it is updated every 5 years. The 2005 responsibility ratio for a country accounts for emission reductions from 2011 to 2015, the 2010 ratio is used for reductions between 2016 and 2020, and so on. By using the 2005 ratio to define reductions from 2011 to 2015, the methodology reflects the lag between actual emissions and their official inventory and reporting.

Unlike the Per Capita Convergence scenario, this allowance allocation requires all countries to reduce their emissions below BAU by 2030 (Table 9.2). Annex I countries, however, bear a greater responsibility for cumulative emissions and thus have more stringent reductions to achieve. The earlier- and/or more heavily-industrializing countries bear the brunt of the required reduction. Simultaneously, the 2030 emission allowances for the non-Annex I countries are greater than their respective 1990 levels, giving them room to achieve development goals.

Table 9.2. Emission Allowances in the Relative Responsibility Scenario

	BAU	*Relative Responsibility Scenario*	
Countries	2030 Emissions (MtC)	2030 Allowances (MtC)	Reduction (–) or increase (+) in emissions relative to 1990 (%)
Annex I			
United States	1,951	1,269	–5
European Union	1,067	570	–34
Japan	331	236	–19
Australia and New Zealand	158	126	+59
Former Soviet Union	944	654	–32
Other Economies in Transition	282	173	–37
Annex I, all others	210	146	+6
Non-Annex I			
Brazil	226	199	+264
Mexico	183	152	+88
India	1,180	1,078	+557
South Asia, excl. India	179	165	+635
China	2,395	2,101	+223
South Korea	249	220	+239
Southeast Asia	921	830	+289
Africa	716	634	+274
Gulf States	473	408	+194
Non-Annex I, all others	516	439	+151
World	11,981	9,400	+66

Source: Calculated using the POLES model.
Note: See Appendix 9B for the definition of geographic regions.
Abbreviations: Business as usual (BAU), millions of tons of carbon equivalent (MtC).

IV. The Emissions-Intensity Target Scenario

Dynamic targets can be expressed in various forms (see Chapters 5 and 6). This chapter refers to emissions intensity, defined as the ratio of CO_2 emissions to gross domestic product (GDP). This scenario does not demonstrate how intensity targets reduce cost uncertainties (see Chapter 5), but rather illustrates how differentiated commitments can be defined in terms of intensity targets. The scenario is built on country-level targets expressed as relative changes to BAU emissions intensities, rather than absolute changes. If such a scheme were under discussion in the international climate negotiations, absolute levels of emissions intensities would not need

Table 9.3. Emission Allowances in the Emissions-Intensity Target Scenario

	BAU	*Emissions-Intensity Target Scenario*	
Countries	**2030 Emissions (MtC)**	**2030 Allowances (MtC)**	**Reduction (–) or increase (+) in emissions relative to 1990 (%)**
AnnexI			
United States	1,951	1,257	–6
European Union	1,067	584	–33
Japan	331	198	–32
Australia and New Zealand	158	83	+5
Former Soviet Union	944	623	–35
Other Economies in Transition	282	186	–32
Annex I, all others	210	102	–26
Non-Annex I			
Brazil	226	205	+275
Mexico	183	166	+105
India	1,180	1,068	+551
South Asia, excl. India	179	161	+620
China	2,395	2,167	+233
South Korea	249	225	+248
Southeast Asia	921	833	+291
Africa	716	648	+282
Gulf States	473	428	+209
Non-Annex I, all others	516	467	+167
World	11,981	9,400	+66

Source: Calculated using the POLES model.
Note: See Appendix 9B for the definition of geographic regions.
Abbreviations: Business as Usual (BAU), millions of tons of carbon equivalent (MtC).

to be compared across countries, obviating the need to agree on an international measurement unit for GDP (see Chapter 5). GDP could be measured in national currencies, rather than U.S. dollars or by using purchasing power parities.

Many simulation options achieve the 2030 emissions target of 9.4 GtC. For example, Annex I countries could improve emissions intensity by 4 percent annually relative to BAU, allowing non-Annex I countries to follow their BAU paths. However, this does not meet the initial assumption that all countries participate in the mitigation effort.

A simulation that meets this participation criterion is one in which Annex I countries improve their emissions intensity by approximately 2 percent[9] annually from their BAU activities, while non-Annex I countries improve their emissions intensity by 0.5 percent. This would amount to a 34 percent improvement in emissions intensity from BAU levels for Annex I countries, and almost 10 percent for non-Annex I countries by 2030.

Improving emissions intensities by 2 percent yearly relative to their BAU levels would imply an approximate 3 percent annual reduction in carbon intensity for most Annex I countries. Intensity changes required to meet the targets in non-Annex I countries would vary more widely, ranging from a 0.5 percent annual intensity increase to a 2 percent decrease. Still, on average, most countries would have to reduce their emissions intensity by around 1 percent annually.

The emission allowances for each country based on this scenario are outlined in Table 9.3. As in the Relative Responsibility scenario, all countries would have to reduce their emissions compared with their BAU levels by 2030. Annex I countries, except Australia/New Zealand, would need to reduce their emissions below 1990 levels,[10] whereas non-Annex I countries could allow their emissions to grow, but at a lower rate than the BAU path.

V. Comparative Assessment

Although the 2030 global emission budget remains unchanged (9.4 GtC), the three scenarios yield different distributions of emission allowances between Annex I and non-Annex I countries (Table 9.4). Namely, Annex I countries would be allocated one quarter of the total global emission budget under the Per Capita Convergence scenario, whereas they would receive approximately one third of all allowances under the other two scenarios. Given the assumptions adopted for each scenario, in all cases Annex I countries need to achieve a greater reduction in their emissions than non-Annex I countries.

As a result of deeper emission reductions in Annex I countries, and despite faster population growth in non-Annex I countries, the 2030 per capita CO_2 allowance for Annex I countries would diminish in all scenarios relative to 1990, while that of non-Annex I countries would increase (Table 9.5). For example, the European Union's per capita allowance would be about 1.6 to 1.7 tC in any of the scenarios, whereas it was 2.4 tC in 1990 and would be 2.8 tC in the BAU case by 2030. In contrast,

Table 9.4. Distribution of CO_2 Emission Allowances Under Three Allocation Scenarios

	1990	2010	2030	2030 allowances		
	Actual Emissions	Projected Emissions	Projected Emissions (BAU)	Per Capita Convergence Scenario	Relative Responsibility Scenario	Emissions-Intensity Target Scenario
Annex I (%)	69	51	41	26	34	32
Non-Annex I (%)	31	49	59	74	66	68
Total (MtC)	5,679	7,832	11,981	9,400	9,400	9,400

Source: Calculated using the POLES model and ASPEN software.
Abbreviations: Business as usual (BAU), millions of tons of carbon equivalent (MtC).

India's 2030 per capita allowance would range between 0.8 and 1 tC in all scenarios, compared with 0.2 tC in 1990 and 0.8 tC in the 2030 BAU case.

From large divergences in projected per capita emissions in 2010, the three scenarios exhibit a trend toward convergence. This result constitutes evidence that per capita convergence may be achieved through various emission-limitation patterns.

The analysis also compares the three scenarios using emission permit trading (Table 9.6). Based on the respective marginal abatement costs of the various countries, and assuming the market in emission permits has opened by 2030, the model calculates a permit price of $97 per ton of carbon equivalent.[11] Countries with marginal abatement costs higher than the permit price would buy allowances to meet their target, while countries with lower marginal costs would sell allowances up to the level at which their marginal cost equals the permit price. The "trade volume" in Table 9.6 refers to the number of allowances (in millions of tons of carbon equivalent) that would be traded. The "total cost to meet the target" corresponds to the cost of reductions achieved domestically and the value of allowances traded. The total cost at the world level is the same in the three scenarios because the global emission target (the environmental goal by 2030) is the same.

The "gains from trade" represent the costs avoided (or, in some cases, benefits generated) with trading. They stem from the difference between the costs that the countries would bear if they achieved their targets solely by reducing emissions domestically and the cost of meeting their targets using emissions trading ("total cost to meet target" in Table 9.6).

Table 9.5 Per Capita CO_2 Emissions Under Three Allocation Scenarios

tons of carbon per capita

	1990	2010	2030			
				Allowable amount of CO_2 emissions		
	Actual Emissions	Projected Emissions	Projected Emissions BAU	Per Capita Convergence Scenario	Relative Responsi-bility Scenario	Emissions-Intensity Target Scenario
Annex I						
United States	5.1	5.6	5.8	2.6	3.8	3.7
European Union	2.4	2.4	2.8	1.7	1.6	1.6
Japan	2.4	2.4	2.8	1.7	2.0	1.7
Australia and New-Zealand	3.0	3.1	3.8	1.6	3.1	2.0
Former Soviet Union	3.3	2.0	3.2	1.6	2.2	2.1
Other EITs	2.2	1.9	2.5	1.6	1.5	1.6
Annex I, all others	3.5	3.8	4.1	1.7	2.8	2.0
Non-Annex I						
Brazil	0.4	0.6	1.0	1.0	0.9	0.9
Mexico	1.0	1.1	1.4	1.2	1.1	1.2
India	0.2	0.5	0.8	1.0	0.8	0.8
South Asia, excl. India	0.1	0.2	0.3	0.8	0.3	0.3
China	0.6	1.1	1.6	1.2	1.4	1.5
South Korea	1.5	2.8	4.7	1.8	4.2	4.3
Southeast Asia	0.4	0.6	1.2	1.0	1.1	1.1
Africa	0.3	0.3	0.5	0.9	0.5	0.5
Gulf States	1.2	1.6	1.9	1.2	1.6	1.8
Non-Annex I, all others	0.6	0.7	1.0	1.0	0.9	0.9
World	1.1	1.1	1.5	1.2	1.2	1.2

Source: Calculated using the POLES model and ASPEN software.
Note: See Appendix 9B for the definition of geographic regions.
Abbreviations: Business as usual (BAU), economies in transition (EITs).

Table 9.6 shows that, in general, Annex I countries would buy allowances, while non-Annex I countries are the sellers in each scenario. The total cost to meet the target is typically higher in Annex I countries than in non-Annex I countries, be it in absolute terms or expressed as a percentage of GDP. Annex I countries are required to undertake more stringent reductions and frequently face higher marginal abatement costs than non-Annex I countries.

These generalizations do not apply to all countries in each group. A few exceptions may be noted. In the Relative Responsibility scenario, the Annex I countries of Australia/New Zealand would be sellers. In the Per Capita Convergence scenario, among non-Annex I countries, the Gulf

Table 9.6. Comparison of Economic Impact of CO_2 Emissions Trading Under Three Allocation Scenarios, 2030

Permit price: $ 97	Per capita convergence scenario				Relative responsibility scenario				Emissions-intensity target scenario			
	Trade Volume	Total cost to meet target		Gains from trade	Trade Volume	Total cost to meet target		Gains from trade	Trade Volume	Total cost to meet target		Gains from trade
	MtC	Billion $	% GDP	Billion $	MtC	Billion $	% GDP	Billion $	MtC	Billion $	% GDP	Billion $
Annex I												
United States	557	75.1	0.48	81.9	165	36.9	0.24	4.9	177	38.1	0.24	5.8
European Union	269	33.1	0.21	47.1	297	35.9	0.23	60.7	283	34.5	0.22	53.6
Japan	100	11.0	0.24	26.9	66	7.7	0.17	11.4	104	11.4	0.24	28.6
Australia and New Zealand	41	6.1	0.49	6.3	–21	0.1	0.01	0.6	22	4.3	0.35	1.3
Former Soviet Union	281	35.7	1.27	66.9	93	17.4	0.62	5.0	125	20.5	0.73	9.8
Other Economies in Transition	48	7.0	0.40	3.6	56	7.7	0.44	5.0	43	6.5	0.37	2.8
Annex I, all others	82	9.5	0.47	19.9	26	4.1	0.20	1.3	70	8.3	0.41	14.6
Non–Annex I												
Brazil	–28	–1.6	–0.06	1.6	4	1.5	0.06	0.0	–2	0.9	0.03	0.0
Mexico	–6	0.9	0.04	0.1	–2	1.3	0.06	0.0	–15	0.0	0.00	0.4
India	–415	–28.6	–0.35	28.6	–161	–3.9	–0.05	5.5	–150	–2.8	–0.03	4.8
South Asia, excl. India	–351	–32.9	–1.89	32.9	–15	–0.2	–0.01	0.5	–12	0.1	0.01	0.3
China	–42	24.2	0.13	0.2	–365	–7.2	–0.04	12.2	–431	–13.6	–0.07	16.5
South Korea	110	12.5	0.67	31.6	–14	0.4	0.02	0.3	–19	–0.1	0.00	0.5
Southeast Asia	–65	2.3	0.03	1.4	–106	–1.7	–0.02	3.4	–109	–2.0	–0.03	3.6
Africa	–613	–55.6	–1.38	55.6	–17	2.4	0.06	0.2	–30	1.1	0.03	0.6
Gulf States	97	13.0	0.63	13.6	–17	1.9	0.09	0.2	–37	–0.1	–0.01	1.0
Non–Annex I, all others	–66	–3.5	–0.05	3.5	11	3.9	0.06	0.1	–17	1.2	0.02	0.3
World	(1586)	108.2	0.11	421.6	(717)	108.2	0.11	111.4	(824)	108.2	0.11	144.7

Source: Calculated using ASPEN software. **Notes:** Monetary values are expressed in constant 1995 dollars, using purchasing power parities. A negative sign in the volume of trade corresponds to permit sales. A negative sign in the total cost corresponds to a benefit. See Appendix 9B for the definition of geographic regions. **Abbreviations:** Millions of tons of carbon equivalent (MtC), gross domestic product (GDP).

States and South Korea would be permit buyers and have the highest cost relative to their GDP.

Interestingly, both the Relative Responsibility and the Emissions-Intensity Target scenarios display similar gains from trading. The Per Capita Convergence scenario would induce the highest global volume of trade in emission allowances and therefore the greatest financial gains from trading. This is mainly because emission reductions required from Annex I (and some non-Annex I) countries are higher in this scenario than in the others. As the permit price is the same in all three scenarios, the level of domestic reductions is the same.[12] Thus, on the one hand, in the Per Capita Convergence scenario, these countries would buy more allowances on the permit market to achieve their more stringent reductions. On the other hand, some non-Annex I countries would have opportunities to sell surplus allowances (those beyond their BAU emission projections), thus providing the allowances needed by the buyers.

These countries with surplus allowances gain a net benefit compared to a situation in which there is no climate change mitigation action (that is, their "total cost to meet target" is negative). In the Per Capita Convergence scenario, trading would lead to important monetary transfers to some non-Annex I countries that are less well off (South Asia and Africa). These results reaffirm Aslam's assertion in Chapter 8 that the "inclusion of trading is... deemed essential for the relative success and appeal of the [per capita convergence] approach." However, the comparative results show that emissions trading is important to reducing costs in all three scenarios examined.

Finally, Table 9.6 also shows that the three differentiation scenarios yield varying abatement costs across countries. Some countries would incur the lowest cost in the Per Capita Convergence scenario, others in the Relative Responsibility scenario, and the remaining in the Emissions-Intensity Target scenario.

Conclusion

This chapter illustrates how a few emission allocation proposals may be formalized. Based on the assumption that emission reductions from BAU are needed to reach a predetermined CO_2 concentration level, it shows how the various proposals can help in the near term to meet a long-term environmental outcome. The same emission reduction target from a BAU level is set to allow comparisons among the scenarios.

In the three scenarios considered, Annex I countries must make larger emission reductions than non-Annex I countries. In the Per Capita Convergence and the Relative Responsibility scenarios, this is due to the current and historically higher levels of emissions of Annex I countries, while in the Emissions-Intensity Target scenario, it is a result of how the model is formulated. Non-Annex I countries receive a larger portion of the carbon budget in all scenarios, enabling these countries to continue developing.

Emissions trading reduces abatement costs, allowing all countries to benefit from trading. In general, Annex I countries are the buyers and non-Annex I countries are the sellers. In terms of the costs (relative to GDP) incurred to meet the target, costs are higher in Annex I countries than in non-Annex I countries.

It must be remembered that these results are dependent on the assumptions adopted. First, they rely on the BAU assumptions of the model used, as well as on the model structure. Second, they relate only to CO_2 emissions from the energy sector. The overall picture would inevitably change if all GHGs were included. Furthermore, trade is assumed to occur in a perfectly competitive market at the international level. However, achieving a well-functioning trading market might be challenging (Baumert et al. 2002). The costs presented are only for emission reductions and do not include transaction costs from trading or costs associated with reporting and monitoring emission inventories. Finally, the regional coverage adopted in this chapter (see Appendix 9B) may hide large disparities across the countries that compose each region. For example, within Africa, the economic structures and the emission levels of oil-producing countries, such as Algeria and Libya, are very different from those of sub-Saharan countries. Adopting the more extensive regional disaggregation of the POLES model (38 countries/regions) would partially overcome this drawback.[13] Still, it may be necessary to undertake country-specific analyses for those countries not individually covered by the model, such as the analysis done for South Africa by Winkler et al. (2001).

Some issues were intentionally omitted from the analysis in this chapter; for example, the practical implementation of any of the scenarios or how countries would meet domestic emission reductions internally. The assumption is that these factors would not have changed the outcome of our analysis.

The results show that the three scenarios examined yield varying abatement costs across countries. This has important political implications, as countries tend to be more prone to accept the solution that is the least costly for them. The analysis provides information to countries on the order of magnitude of emission reductions and the associated costs, depending on the scenario used. This information may be helpful to countries in shaping their own negotiating positions. For some countries, none of the scenarios considered may be acceptable from a sustainability standpoint. Of course, the whole spectrum of options for global participation in a climate change mitigation effort is much wider than those analyzed here. Approaches such as Sustainable Development Policies and Measures or the Sectoral Clean Development Mechanism (addressed in Chapters 3 and 4) may also contribute to emission reductions in some countries, while bringing them other development benefits.

This analysis could be further developed to explore the consequences of exempting some countries from emission limitations. The exemption could apply to those countries whose emissions and GDP per capita are very low and that do not significantly contribute to the build-up of global CO_2 emissions.

Appendix 9A

The POLES Model: Prospective Outlook on Long-Term Energy Systems

The POLES model was developed at Institut d'Economie et de Politique de l'Energie in Grenoble, France, under research programs funded by the European Union. Operational since 1997, it has been used for policy analyses by various European Commission's-Directorates General (e.g. DG-Research, DG-Environment) and by the French Ministry of Environment.

POLES is a world simulation model for the energy sector. It works in a year-by-year recursive simulation and partial equilibrium framework, with endogenous international energy prices and lagged adjustments of supply and demand by world region. GDP and population are the main exogenous variables.

POLES model structure

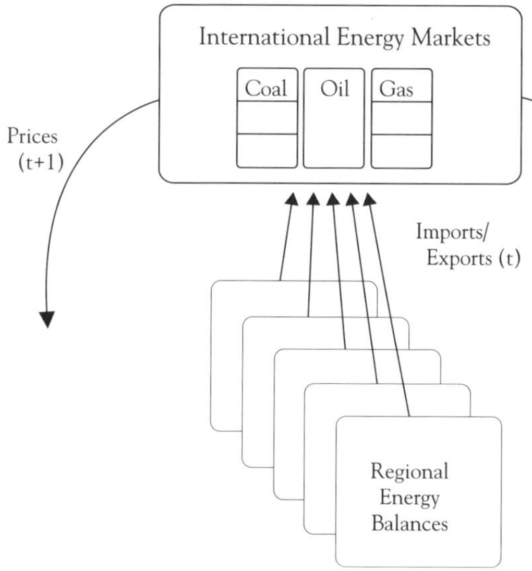

In the current geographic disaggregation of the model, the world is divided into 38 countries or regions. For each region, the model articulates four main modules dealing with the following:

- Final energy demand by key sectors
- New and renewable energy technologies
- The conventional energy and electricity transformation system
- Fossil fuel supply

The main outputs are the following:

- Detailed world energy outlooks to 2030, with demand, supply, and price projections by global regions.
- CO_2 emission marginal abatement cost curves by region, and emissions trading systems analyses.
- Technology improvement scenarios—exogenous or with endogenous features—and analyses of the value of technological progress in the context of CO_2 abatement policies.

The main advantages of the POLES model rely on the high disaggregation levels of energy demand sectors, energy technologies, and geographic regions. The detailed representation of the energy sector allows it to endogenously capture the various changes, such as the development and implementation of economically efficient new technologies. The geographic breakout delivers detailed insights on energy variables and CO_2-related emissions for many countries. This feature allows the model to better illustrate the challenges of many countries, and may be useful in the course of the climate negotiations.

See European Commission (1996) for a comprehensive description of the POLES model.

Appendix 9B

Regional Breakdown

The regional breakdown indicates how the 38 countries or regions of the POLES model are aggregated for the purpose of this chapter. A "+" sign means that the countries are grouped to constitute a single element.

United States: United States of America.
European Union: Includes Austria, Belgium+Luxemburg, Denmark, Finland, France, Germany, Greece, Ireland, Italy, the Netherlands, Portugal, Spain, Sweden, and the United Kingdom.
Japan
Australia and New Zealand
Former Soviet Union
Other Economies in Transition: Includes Poland+Hungary+the Czech Republic+Slovenia, Rest Eastern Europe.
Annex I, all others: Includes Canada, Rest Western Europe.
Brazil
Mexico
India
South Asia, excluding India
China
South Korea
Southeast Asia
Africa: Includes North Africa non-OPEC, North Africa OPEC, Egypt, Sub-Saharan Africa.
Gulf States: Includes OPEC countries in the Persian Gulf.
Non-Annex I, all others: Includes Rest Central America, Rest South America, Rest Middle East, and Turkey.

Notes

1. "Countries" is used in this chapter as a generic term covering individual countries as well as geographic regions.
2. The Analyse des Systèmes de Permis d'Emissions Négociables (ASPEN) software was also developed at Institut d'Economie et de Politique de l'Energie (IEPE) in Grenoble, France. It computes regional CO_2 emission allowances according to a given differentiation approach. It then uses the POLES marginal abatement costs to compute the emission permit price and trade flows for any configuration of the emission permit market.
3. Thus, net CO_2 emissions resulting from land-use change activities as well as the emissions of other GHGs are not considered.
4. As in any model, the BAU case is built on many assumptions. For example, the projected GDP annual growth rate of each country, determined exogenously, leads to a worldwide average yearly GDP increase of 2.9 percent between 2010 and 2030.
5. South Korea would obviously have a tremendous challenge.
6. The data source used in the chapter is the World Resources Institute, gathered from EIA (2002b) and Marland et al. (2000).
7. This is because the method used in this scenario does not account for the decay of emissions over time. See Chapter 7.
8. As stated in the previous section, the budgets are defined so that from 2010 to 2030, the global emissions grow linearly from 7.8 to 9.4 GtC.
9. The precise rate is 2.059 percent.
10. Australia/New Zealand and the United States would be in a comparable situation by 2030. Their reductions relative to 1990 levels would not be as stringent as for the other Annex 1 countries because their assumed departure point in 2010 is above 1990 levels (as opposed to the other Annex 1 countries).
11. The permit price may seem high. This may be explained by the fairly high volume of emission reductions.
12. Buying countries will reduce their emissions domestically as long as the marginal cost of these reductions is lower than or equal to the permit price. They will turn to the international permit market for those additional reductions whose marginal cost, if taken domestically, would exceed the permit price.
13. Regional disaggregation results obtained using the POLES model are not presented in the chapter, but can be provided upon request.

10. CONCLUSION:
Building an Effective and Fair Climate Protection Architecture

Kevin A. Baumert and Silvia Llosa

As the Kyoto Protocol comes into force, the climate protection debate will focus increasingly on improving and expanding the regime's architecture in the era beyond Kyoto's first commitment period. Official negotiations could begin between 2003 and 2006. The approaches to climate protection analyzed in this volume suggest a wide variety of future possibilities for the evolution of the climate regime, some of which are examined here for the first time.

Crafting future agreements will require governments to converge on a mutually agreeable course of action, as decisions are made by consensus under the Climate Convention. Yet, as several studies in this volume reflect, governments act in their own interests, even in trying to address a global problem such as climate change. Aslam illustrates in Chapter 8 that climate protection proposals tend to primarily benefit their proponents, rather than strive to meet the needs of all countries. Reconciling these realities with the need to slow and eventually reduce global greenhouse gas (GHG) emissions over the coming decades is a paramount challenge. What might the features of a successful climate protection architecture look like? Is there a winning combination of elements among the options and approaches outlined in this volume? This chapter responds to these questions.

I. Interests and Architecture

To begin, none of the approaches examined can satisfy the interests and concerns of all countries. Table 10.1 amply illustrates this, presenting the advantages and disadvantages of various approaches as catalogued by the authors of this volume's respective chapters. The results should not be surprising. Disparate economic conditions, demographic profiles, values, and other unique factors prevent any single approach from suiting the

Table 10.1. Summary of Major Advantages and Disadvantages of Selected Climate Protection Approaches

Approach	Major Advantages	Disadvantages/ Challenges
Kyoto Protocol-Style Targets *Extending fixed targets to developing countries with links to flexibility and accountability mechanisms*	▪ Familiarity and simplicity ▪ Advance knowledge of environmental benefits ▪ Flexibility in implementation ▪ Respects national circumstances	▪ Limited political acceptability ▪ Data requirements ▪ Incentives to establish weak targets
Sustainable Development Policies and Measures (SD-PAMs) *Voluntary action oriented around sustainable development*	▪ Builds on national sustainable development priorities ▪ Respects national circumstances ▪ Easily integrated into Kyoto Protocol ▪ No emissions cap	▪ Ensuring action and accountability ▪ Measuring climate benefits ▪ Financing
Sector-Clean Development Mechanism (Sector-CDM) *Sector-wide market mechanism*	▪ Familiarity and compatibility with the Protocol ▪ Development benefits ▪ Rests on the polluter pays principle ▪ Gradual capacity building ▪ Cost-effectiveness ▪ No emissions cap	▪ Relies on Annex I investment ▪ Technical requirements and capacity ▪ National coordination effort ▪ Political opposition
Dual-Intensity Targets *Two dynamic targets with links to flexibility and accountability mechanisms*	▪ Reduced economic uncertainty in establishing targets ▪ Reduces risk of hot air targets ▪ Potentially easier to agree on dual targets	▪ Data requirements ▪ Complexity ▪ Interactions with international emissions trading ▪ Lack of environmental certainty
Adaptation of Brazilian Proposal, as suggested in Chapter 7 *Fixed target, global allocation scheme with links to flexibility and accountability mechanisms*	▪ Procedural fairness and simplicity ▪ Science-driven ▪ Rests on established principles ▪ Compatibility with Kyoto Protocol mechanisms ▪ Rewards of early developing country action	▪ Data requirements ▪ Limited global acceptability ▪ Limited flexibility for varying country circumstances
Per Capita-Based Entitlements *Fixed targets, global allocation scheme with links to flexibility and accountability mechanisms*	▪ Procedural fairness and simplicity ▪ Strong ethical basis ▪ Enhances cost-effectiveness through global trading ▪ Incentives for developing country participation ▪ Amalgamates well with the Kyoto architecture	▪ Limited global acceptability ▪ Limited flexibility for varying country circumstances ▪ High dependence on trading for success

circumstances and interests of all countries. Chapters 8 and 9, for example, show that allocation approaches, such as those based on a relative responsibility or per capita principle, tend to distribute benefits and burdens in a way that some countries might find unacceptable, at least in the near term.

The differences in national circumstances and interests are evident in this volume's case studies on Sustainable Development Policies and Measures (SD-PAMs), Sector-Clean Development Mechanism (Sector-CDM), and dual-intensity targets. South Africa, for instance, would likely fare poorly if emission-limitations commitments were determined through relative responsibility or per capita-based differentiation principles. However, given its social and economic situation—including political commitments to sustainable development—South Africa might be favorably inclined toward an approach such as SD-PAMs. Mexico's commitment to improving local air pollution—as well as the high concentration of emissions in the capital area—makes Sector-CDM attractive. South Korea has a rapidly growing economy where future emissions are highly uncertain; hence, the suitability of dual-intensity targets.

Avoiding false comparisons, however, is important. Although this volume analyzes different approaches side by side, none of them articulates a complete architecture.[1] Rather, the approaches generally depict one or a few elements that could form part of a broader near- or long-term climate protection architecture. Fixed and dual-dynamic targets (Chapters 2 and 5), for example, address the form, or type, of commitment that governments might adopt, but not how that commitment might be differentiated from other countries' commitments. Furthermore, some approaches also vary by the time frame under which they could be made operational. Thus, a coherent architecture is not a matter of picking one of the approaches examined in this volume, just as a homebuilder would not want to choose between using doors and windows. Table 10.2 matches the approaches examined in this volume with their respective, sometimes overlapping, architectural elements. These elements were described in detail in Chapter 1.

Likewise, the approaches examined attempt, in most cases, to address different problems or challenges. For example, for many developing countries, climate protection is not a priority. Thus, sustainable development might be a more compelling issue around which to organize action, as illustrated in Chapter 3 on SD-PAMs. Table 10.3 shows the different challenges that the approaches seek to address, including investment needs (Sector-CDM), the reduction of cost uncertainties (dual-intensity targets), procedural fairness, and differentiation of commitments (principle-based allocation approaches). Each one of the approaches examined in this volume is relevant for thinking about how best to overcome particular obstacles that are important to different countries.

Table 10.2. Elements of a Climate Protection Architecture Found in Selected Approaches

Element of Architecture	Options and Examples (chapter number)
Legal Nature of Commitment	1. Binding: emission targets (2, 5, 7, 8) 2. Non-binding: SD-PAMs (3); Sector-CDM (4); targets (2, 5)
Type of Greenhouse-Gas Limitation Commitment	1. Fixed target: Kyoto Protocol (2); Brazilian Proposal (7); per capita entitlements (8) 2. Dynamic target: Dual-intensity targets (5); Argentine target (6) 3. Dual target: Dual-intensity targets (5) 4. Sustainable development: SD-PAMs (3)
Coverage and Scope of Actions	1. Project: Kyoto Protocol's CDM (2) 2. Sector: Sector-CDM (4) 3. National: Targets (2, 5) 4. Global: Responsibility-based targets (7); per capita-based entitlements (8)
Approach to Differentiating Commitments	1. Pledge-based: fixed (2) and dual-intensity (5) targets; SD-PAMs (3) 2. Principle-based: Brazilian Proposal (7); per capital entitlements (8)
Financial and Other Commitments	Funding for implementation: SD-PAMs (3); implicit in other approaches
Use of Market-Based Mechanisms	1. Project or sector-based trading: Kyoto Protocol (2); SD-PAMs (3); Sector-CDM (4) 2. International emissions trading: Possible with fixed (2) and dual-intensity targets (5); Brazilian Proposal (7); per capital entitlements (8)

Notes: There are other elements of the architecture not shown above, such as *environmental objective* and *gases* covered, where all of these approaches are flexible.
Abbreviations: Sustainable Development Policies and Measure (SD-PAMs), Clean Development Mechanism (CDM).

II. Designing a Menu of Near-Term Options

An overarching lesson emerging from this volume is that the climate protection architecture can be designed to accommodate multiple, sometimes conflicting, interests and national circumstances. In this regard, governments might consider adopting several options for country participation in the near term. Some of the approaches may operate simultaneously in a way that supports a menu-based strategy. Such a strategy could at once address the needs of many different countries, but through different channels of participation.

For example, the SD-PAMs and Sector-CDM could be mutually reinforcing mechanisms. The former provides a platform for promoting cleaner development, while the latter could encourage investment in specific sus-

Table 10.3. Key Challenges Addressed by Selected Climate Protection Approaches

Approach	Major Challenge(s) Addressed
Fixed, Kyoto Protocol-Style Targets (Chapter 2)	Ensuring environmental certainty
Sustainable Development Policies and Measures (Chapter 3)	Promoting sustainable development in developing countries
Sector-Clean Development Mechanism (Chapter 4)	Financing emission reductions and sustainable development in developing countries
Dual-Intensity Targets (Chapter 5)	Reducing cost uncertainties
Brazilian Proposal (Chapter 7) Per Capita Entitlements (Chapter 8)	Promoting procedural fairness; differentiating greenhouse-gas limitation commitments at the country level

tainable development policies that also deliver climate benefits. Similarly, fixed and dynamic targets could operate in different countries simultaneously. Dual-intensity targets (one kind of dynamic target) might provide a safety valve for some middle-income countries, while fixed targets might be best suited for mature, industrialized economies. Both types of targets could support the use of cost-saving market mechanisms. To be successful, developing countries would need to play an active role in defining the contents of the menu, rather than simply reacting to the proposals of industrialized countries.

Such a menu-based approach would have several advantages. First, it may be the best hope of harnessing the limited political will that currently exists for climate protection. A menu of multiple options might comprise various items and choices through which countries could exercise their political will. A single option (e.g., fixed targets) might be too limiting and, consequently, the flexibility of multiple options might mobilize maximum emission reductions.

Second, a system featuring multiple options may enable Parties to gain experience, capacity, and added confidence to support greater long-term action on climate change. It might take 100 or more years to achieve the Climate Convention's ultimate objective of preventing dangerous climate change. Countries need to improve their abilities to measure, report, and manage GHGs if this objective is to be reached. SD-PAMs, Sector-CDM, and dual-intensity targets (with no compliance target, for example) could help build experience and capacity in developing countries through concrete action. A modest start could yield even larger long-term payoffs.

III. The Need for a Principled, Long-Term Framework

The near-term trust-building described above is especially important because, in all likelihood, a menu-based system is unlikely to be sufficient to address climate change over the long term. Rather, achieving the Convention's objective may require a principle-based framework, as exemplified in the Brazilian Proposal and per capita-based approaches. Several interrelated factors suggest this long-term need.

The first factor is *procedural equity*, which will become an increasingly important issue as the climate regime expands to cover more countries under GHG limitation commitments. A system of pledged, negotiated commitments (exemplified in the present regime and in the approaches discussed in Chapters 3 and 5) might place some developing countries at an inherent disadvantage because of the sheer bargaining power of the industrialized countries, and even other developing countries. The significance of this disadvantage depends, of course, on what is being bargained for. If the subject is legally binding emission limits, then many developing countries may justifiably seek a more objective framework within which to negotiate those commitments. Otherwise, doubts may persist over whether the process can deliver fair results. Even though the Brazilian Proposal and the per capita approaches do not represent universally agreed-on principles, they are nonetheless the kind of approaches that might help level the playing field and combat the underlying power structure of the international order.

A second factor is *complexity*. As pointed out by Depledge (Chapter 2), the climate regime is already highly complex. Including multiple options for participation would undoubtedly increase this complexity. Even a single approach, such as dual-intensity targets, could be complicated, perhaps characterized by country-specific indexes for target adjustments. Whether the Convention process could cope with the added complexity would depend on the time frame for reaching agreement, the capacities of the negotiating parties, and the number of countries seeking to exercise the various options. Complexity, when combined with the weak negotiating capacity of many governments, can lead to defensive negotiating postures and a culture of mistrust.

At some point, complexity can become the enemy of *environmental effectiveness*. Complexity reduces transparency and allows countries to conceal weak negotiating positions in a shroud of numbers, terminology, and other obscurities that are beyond the understanding of all but a few insiders. This, in turn, reduces the ability of governments and civil society to expose weak negotiating positions. This phenomenon is already evident in the Kyoto Protocol. The complexity of the accounting rules for emissions and absorptions from the land-use change and forestry sectors, for example, prevents the media and the public from exercising scrutiny over government positions.

A principled, long-term framework might include a more definite overall objective (e.g., aiming for atmospheric stabilization of CO_2 at 450 ppmv), a system for differentiating GHG limitation commitments, and more robust financial commitments, among others. Such a framework is not an impossible goal, and it is one that governments and civil society should promote and analysts should explore. However, such a long-term framework would need to overcome some of the persistent barriers identified in this volume, especially with respect to differentiating GHG limitation commitments. Grubb et al. (1999) describe such proposals as offering "a logical, top-down and long-term resolution in the context of a political process that is inherently illogical, bottom-up and mostly concerned with the current or next round of commitments." Countries tend to adopt near-term commitments that they deem economically acceptable. Some overarching principles may even conflict with other established principles, such as the need to take into account specific national circumstances, a principle that is firmly embedded in the Climate Convention. Furthermore, various principles and formulas—sometimes invoked under the guise of equity or fairness—are often designed to serve the interests of particular countries.[2]

Thus, an important area of future research is to explore how a long-term framework might better take into account national circumstances and thus gain wider acceptability. Two preliminary ideas emerge from this project. First, as Aslam suggests in Chapter 8, a scheme such as per capita-based entitlements could have two components—a "fixed," equal per capita component and a second, "variable" one that accounts for national circumstances or incorporates other principles. Countries could even invoke the fairness principles they endorse in articulating the variable portion of their emission target. Second, other elements of the architecture could compensate for commitments that are overly burdensome for some countries. Technology transfer, clean energy funds, or other financing provisions could be created (or existing systems expanded) to assist countries in meeting their commitments. This approach was used in the Montreal Protocol's architecture for protecting the stratospheric ozone layer. The Montreal Protocol adopted uniform reductions of ozone-depleting substances from historical levels (although with a grace period for developing-country groups) and then created a multilateral fund to finance those phase-outs in developing countries. The acceptability was due not to the targets per se but to the mixture of the targets, a grace period, and a credible financial mechanism.

Overall, the preceding analysis suggests a two-track strategy. A near-term component might leave open several options for country participation. Yet a parallel strategy that aims at a more coherent and principled long-term framework for climate protection should accompany this strategy. For example, an agreement on a near-term menu approach might be accompanied by a timetable for negotiating a principled, longer-term framework. In both tracks, promoting equity and environmental effectiveness will require using a suite of elements in the climate protection architecture, including financial commitments for investment in clean energy or compensation for adverse climate impacts suffered by the poor. Here, creativity and innovation are possible and indeed may be necessary.

IV. Lessons for Building an Effective and Fair Climate Protection Architecture

Key lessons that emerge from the above discussion and the entire volume of studies include the following:

The current regime provides a solid foundation on which to build.

Both the Convention and the Kyoto Protocol include the necessary foundations for further developing the climate regime. These include provisions both for developing countries to individually accede to fixed emission targets and for the launch of a comprehensive negotiating round, whose results could encompass any of the options discussed in this volume (and others not examined here). Even if new options for GHG limitation commitments are adopted in a new negotiating round, the Protocol's reporting and review systems, as well as its flexibility mechanisms, provide a sound basis for future development.

The climate regime needs stronger leadership and U.S. participation.

The weakness of industrialized country leadership—especially the United States, which has repudiated the Kyoto Protocol—is the greatest barrier to fruitful North-South cooperation. Without greater action by the United States, the approaches examined within this volume are unlikely to be viable. Specifically, Sector-CDM requires deeper emission cuts from industrialized countries to generate demand for Sector-CDM projects in developing countries. Implementation of SD-PAMs would depend on industrialized-country funding. Emission targets (fixed or dynamic) are not remotely feasible in developing countries without a prior demonstration of industrialized country leadership, as called for in the Climate Convention. Generally, the actions of the United States illustrate the broader limitations of the international legal order. There are simply no tools readily available to force recalcitrant countries into more climate-friendly behavior. In this way, international negotiations are no different than geology: time and pressure are needed to drive change.

While industrialized-country leadership is needed, this requirement should not preclude developing countries from exercising their own leadership. In fact, there is a route out of the current stalemate—the United States, other industrialized countries, and some high-emitting developing countries might participate together in the next round of mitigation commitments, albeit under a differentiated system in which industrialized countries make the most substantial commitments. Waiting for the United States to re-engage may be in no one's interest. There is an opportunity for developing countries to help define the terms of their future participation, to shape their involvement in a way that promotes their development objectives while slowing emissions growth. In fact, the ideas and insights in this volume are largely those of analysts from developing countries, and many of the approaches examined can be shaped in a way that would support development objectives.

Information systems and capacity in developing countries need strengthening.

Another major barrier to expanding the climate regime to include GHG limitation commitments in developing countries is the paucity and poor quality of climate change-related data and information. Sound decisions require a strong information base. As Depledge pointed out in Chapter 2, many developing countries have not yet submitted their first national reports under the Convention. This highlights the need for a massive international effort to improve the collection and analysis of emissions data in developing countries. The weak information base also extends to national mitigation and adaptation assessments. The sharing of benefits and burdens is central to the negotiation of a climate protection architecture and specific national commitments. Yet, it is difficult to talk about sharing burdens and benefits without more information about their actual magnitude. Progress on information systems will require capacity building and financial and technical assistance.

For Argentina (Chapter 6), the formulation of a voluntary target showed the importance of domestic capacity in economic modeling and emissions inventories, among other areas. Estimates by outside experts are unlikely to be sufficient to shape important national decisions that might entail future obligations and associated costs. Approaches such as the Sector-CDM or non-binding targets might support capacity building at the national level. Targeted and effective North-South transfers (financial resources, skills, etc.) are clearly needed to support future actions by developing countries. However, successful capacity building also requires developing countries to make political commitments that mobilize and enhance the existing expertise, as Chapter 6 makes clear.

Successful implementation requires domestic buy-in and acceptance. Argentina's experience with a voluntary target highlights the importance of national buy-in for internationally agreed measures. Without domestic public understanding of the issue, as well as support and coordination among the multiple sectors involved, implementation of an international commitment may fail. Had Argentina's voluntary commitment been accepted by other countries, the lack of domestic support, coupled with the economic crisis and changes in political leadership, would have made implementation difficult or impossible. Garnering support from different political constituencies, key industries, provincial governments, and the public will improve the viability of any approach. The difficulty of achieving buy-in on a relatively low-priority issue such as climate change highlights the broader challenges of North-South cooperation. A climate proposal's acceptance in a developing country will likely be based on its contribution to domestic priorities. Thus, approaches that can provide such tangible benefits while also assisting in climate protection—such as SD-PAMs and Sector-CDM—may be more likely to find domestic support in the near term.

In the North and the South, governments and civil society have important roles to play in increasing public concern for the issue of climate change and articulating the relationship between climate protection and other important priorities, such as local pollution and poverty. Greater interest in and concern for climate protection in general will help enable the public to make informed contributions to national strategies for climate protection.

Simplicity is a virtue.

Simplicity promotes transparency, fairness, and good-faith negotiating. A drawback of the menu approach described above is that the complexity of the negotiations could become overwhelming. There is an apparent trade-off between simplicity and meeting Parties' diverse interests. One hope for limiting the complexity of a menu-based approach is to promote simplicity within each of the menu's options. For example, if countries decide to pursue dual-intensity targets, they should settle on a few transparent options for target adjustment. Otherwise, each country will formulate its own target, devising formulas that few others will understand, similar to the case of Argentina (see Chapter 6).

Another possible way to promote simplicity would be to establish a trigger (or multiple triggers) that would signal the need for a country to take greater action. Under any number of triggers—based on capability, responsibility, or even current GHG emissions—most developing countries would probably be exempt from serious actions. Establishing these kinds of participation thresholds could help focus attention on the relatively few countries that are large current or future contributors to climate change. As pointed out in Chapter 1, many developing countries have made very little contribution to the buildup of GHGs in the atmosphere.

A two-track future approach might best balance interests, simplicity, and fairness.

Different approaches to climate protection tend to address one or several of Parties' key concerns or interests, such as the need to reduce uncertainty, promote investment, account for national circumstances, establish procedural fairness, and differentiate GHG limitation commitments across countries. Through different elements of the architecture, many of these interests can be accommodated simultaneously. As described earlier in this chapter, combining some of the approaches (e.g., different kinds of targets, Sector-CDM, and SD-PAMs) could help constitute a near-term strategy that responds to the diverse interests and circumstances of Climate Convention Parties. One element of the architecture—financial mechanisms (e.g., funds for technology transfer, clean energy, and compensation)—will be critical for acceptable compromises and might be coupled with emission targets or other commitments.

Over the longer term, complexity and procedural equity suggest the need for a simpler, more coherent framework. This framework could be explored and developed in parallel with a near-term response strategy.

However, no such long-term framework exists (although the Brazilian Proposal is being discussed officially), and it is uncertain what an acceptable one might look like. This constitutes an important area for future research. The most important prospective element of such a framework might be greater clarity on a long-term environmental goal. Other important elements include a framework for shaping emission limits or other commitments for countries with diverse national circumstances and a financial transfer mechanism that addresses mitigation and adaptation needs. To the extent that countries adopt near-term actions that build confidence and trust, the viability of such a long-term climate protection strategy increases.

From the vantage point of the 1992 Earth Summit in Rio de Janeiro, where the Climate Convention was born, one might have looked ahead 10 years and imaged a relatively simple evolution of the climate change regime. Such an evolution might have been similar to other multilateral environmental agreements, perhaps characterized by relatively fixed emission reductions and financial mechanisms. After all, even by 1992, governments had a fair bit of experience and success in dealing with transboundary air pollution and ozone-depleting gases.

Such a view seems naïve today. Climate change is a highly complex issue, characterized by pervasive scientific and economic uncertainties, long time horizons, potentially irreversible change, and a common atmospheric resource shared by 6 billion people and nearly 200 countries. Likewise, climate change mitigation is tethered to economic development and human welfare issues as well as entrenched political interests. Although many of these characteristics of the problem were known 10 years ago, today we have a better sense of the limited capacity of national and global institutions to respond effectively to the challenge. Because today's institutions are relatively poorly equipped to deal with a problem of this scale, creativity and innovation may be needed in designing future solutions. This might be especially true in the near term, where building confidence and experience with emission reductions might pave the way for more effective longer-term solutions. This is the primary reason for examining a diversity of options for a pragmatic, fair, and environmentally effective climate future.

Notes

1. The exception, of course, is Chapter 2, which describes the current climate protection regime.
2. An interesting example of how principles and formulas are invoked to serve country interests can be found in the recent World Trade Organization (WTO) debate on agricultural subsidies (Blustein 2002). The United States has made a proposal that industrialized countries reduce their "trade-distorting" agricultural subsidies under the WTO according to a particular formula (specifically, to 5 percent of a country's farm output). This formula would require the United States to reduce its subsidies from $19 to $10 billion per year, while the European Union (EU) and Japan would be required to cut subsidies from $60 to $12 billion and from $33 to $4 billion, respectively. According to U.S. Trade Representative Robert Zoellick, "Those with the highest barriers and subsidies must cut the most, as is only *fair*" (emphasis added). This formula is not considered "fair" by the EU in this context. Likewise, a similar approach applied in the context of reducing GHG emissions ("those with the highest, cut the most") might not be considered fair by the United States.

Bibliography

ANC (African National Congress). 1994. *The Reconstruction and Development Programme: A Policy Framework.* Johannesburg: Umanyano.

Agarwal, A., S. Narain, and A. Sharma, eds. 1999. "A Boiling Point," in *Green Politics - Global Environmental Negotiations I*. New Delhi: Centre for Science and Environment.

Agarwal, A. and S. Narain. 1991. *Global Warming in an Unequal World.* New Delhi: Centre for Science and Environment.

Albin, C. 2002 forthcoming. "Getting to Fairness Matters: Negotiations over Public Goods," in *Providing Global Public Goods: Managing Globalization*, I. Kaul, ed. New York: Oxford University Press.

Argentine Republic. 1999. *Revision of the First National Communication*. Buenos Aires: Secretariat for Natural Resources and Sustainable Development.

Aslam, M.A. 2001. "International GHG Emission Trading and Issues Related to Voluntary Participation," in *GHG Market Perspectives: Trade and Investment Implications of the Climate Change Regime*, M.A. Aslam, J. Cozijnsen, M. Stuart, R. Stewart, and P. Sands, eds. Geneva: UNCTAD.

Austin, D., G. Parker, and J. Goldemberg. 1998. *Contributions to Climate Change: Are Conventional Metrics Misleading the Debate?* Washington, DC: World Resources Institute.

Baer, P. et al. 2000. "Equity and Greenhouse Gas Responsibility." *Science* 289 (September 29).

Banuri, T. and J.P. Weyant. 2001. "Setting the Stage: Climate Change and Sustainable Development," in *Climate Change 2001: Mitigation*, B. Metz et al., eds. Contribution of WG III to the Third Assessment Report of the IPCC. Cambridge: Intergovernmental Panel on Climate Change, Cambridge University Press.

Banuri, T. et al. 1996. "Equity and Social Considerations," in *Climate Change 1995: Economic and Social Dimensions of Climate Change*, J.P. Bruce, L. Hoesung, and E.F. Haites, eds. Contribution of WG III to the Second Assessment Report of the IPCC. Cambridge: Cambridge University Press.

Barros, V. and M. Conte Grand. 1999. "El Significado de una Meta Dinámica de Reducción de Gases de Efecto Invernadero: El Caso Argentino." XXXV Annual Meeting of the Argentine Association of Political Economy.

Baumert, K.A., R. Bhandari, and N. Kete. 1999. *What Might A Developing Country Climate Commitment Look Like?* Washington, DC: World Resources Institute.

Baumert, K.A. and N. Kete. 2001. *The U.S., Developing Countries, and Climate Protection: Leadership or Stalemate?* Washington, DC: World Resources Institute.

Baumert, K.A., J. Perkaus, and N. Kete. 2003 forthcoming. "Deflating Great Expectations?: International Emissions Trading and Global Climate Equity." *Climate Policy*.

Baumol, W.J. and W. Oates. 1988. *The Theory of Environmental Policy*. Second Edition. New York: Cambridge University Press.

Berk, M.M. and M.G.J. den Elzen. 2001. "Options for Differentiation of Future Commitments in Climate Policy: How to Realize Timely Participation to Meet Stringent Climate Goals?" *Climate Policy* 1(4): 465—480.

Berk, M.M., J.G. Van Minnen, B. Metz, and W.R. Moomaw. 2001. *Keeping our Options Open: A Strategic Vision on Near-Term Implications of Long-Term Climate Policy Options*. Concise summary of the results of the Global Dialogue of the COOL-project (Climate OptiOns for the Long term). Bilthoven, Netherlands: RIVM.

Bertram G. 1992. "Tradable Emissions Permits and the Control of Greenhouse Gases." *Journal of Development Studies* 28(3).

Bertram G., R.J. Stephens, and C.C. Wallace. 1989. "The Relevance of Economic Instruments in Tackling the Greenhouse Effect." Report to New Zealand Ministry of the Environment.

Biagini, B., ed. 2000. *Confronting Climate Change: Economic Priorities and Climate Protection in Developing Nations*. Washington, DC: National Environmental Trust and Pelangi.

Bierman, F. 1999. "Justice in the Greenhouse: Perspectives from International Law," in *Fair Weather: Equity Concerns in Climate Change*, F. Toth, ed. London: Earthscan.

Blanchard, O., P. Criqui, A. Kitous, and L. Viguier. 2002 forthcoming. "Combining Efficiency with Equity: A Pragmatic Approach," in *Providing Global Public Goods: Managing Globalization*, I. Kaul, ed. New York: Oxford University Press.

Blanchard, O., P. Criqui, M. Trommetter, and L. Viguier. 2001. *Equity and Efficiency in Climate Change Negotiations:A Scenario for World Emission Entitlements by 2030*. Cahier de Recherche No. 26. Grenoble, France: Institut d'Economie et de Politique de l'Energie.

Blustein, P. 2002. "U.S. Proposes Global Cut in Farm Subsidies, Tariffs." *Washington Post* (July 26, p. E1).

Bohringer, C. and H. Weisch. 2000. "Contraction of Global Carbon Emissions: How Acceptable are Alternative Emission Entitlement Schemes?" ZEW discussion paper.

Borchers, M., et al. 2001. *National Electrification Programme Evaluation: Summary Report*. Evaluation commissioned by the Department of Minerals and Energy and the Development Bank of Southern Africa. Cape Town: University of Cape Town, Energy & Development Research Centre.

Byrne, J., Y-D. Wang, H. Lee, and J-d. Kim. 1998. "An Equity- and Sustainability-Based Policy Response to Global Climate Change." *Energy Policy* 26 (4): 335—43.

Carlson, C. and D. Burtraw. 2000. "SO_2 Control by Electric Utilities: What Are the Gains from Trade?" *Journal of Political Economy* 108(6): 1292—1326.

Castellini, R. 1998. Interview with Raúl Castellini, Argentine delegate to COP4. Online at http://baires.cop4.org/entrevistas/english/castellini.en.html (June 27, 2002).

CCAP (Center for Clean Air Policy). 1998. *Growth Baselines: Reducing Emissions and Increasing Investment in Developing Countries*. Washington, DC.

Clark, A. and J. Mavhungu. 2000. *Promoting Public Benefit Energy-Efficiency Investment in New Power Contexts in South Africa*. Cape Town: University of Cape Town, Energy & Development Research Centre.

Claussen, E. and L. McNeilly. 1998. *Equity and Global Climate Change: The Complex Elements of Fairness*. Arlington, VA: Pew Center on Climate Change.

Corfee-Morlot, J. 2002. "Climate Change: Long-Term Targets and Short-Term Commitments." Revised draft note. OECD/IEA project for the Annex I Expert Group on the UNFCCC. Paris: OECD/IEA.

CSD (Commission on Sustainable Development). 1995. *Indicators of Sustainable Development: Guidelines and Methodologies*. New York: United Nations. Online at: www.un.org/esa/sustdev/indisd/indisd-mg2001.pdf.

CSE (Centre for Science and Environment). 1998. "Climate No Headway." *Down to Earth* Vol. 7, No. 8 (August 15).

Davidson, O. and N. Nakicenovic. 2001. *Linking Climate Change to Sustainable Development*. Nairobi: United Nations Environment Programme.

Davidson, O. and Y. Sokona. 2001. "Energy and Sustainable Development: Key Issues for Africa," in *Proceedings of the High-level Regional Meeting on Energy and Sustainable Development for the Ninth Sessions of the Commission on Sustainable Development*, N. Wamukonya, ed. Roskilde, Denmark: UNEP Collaborating Centre on Energy & Environment.

Davidson, O., L. Tyani, and Y. Afrane-Okese. 2001. *Climate Change and Sustainable Development: Future Perspectives for South Africa*. Cape Town: University of Cape Town, Energy & Development Research Centre.

Davis, M. and N. Wamukonya. 1999. *Capacity-Building in CDM Project Activities: South Africa Interim Report*. Cape Town: University of Cape Town, Energy & Development Research Centre.

Dixon, R.K., ed. 1999. *The UNFCCC Activities Implemented Jointly (AIJ) Pilot: Experiences and Lessons Learned*. Dordrecht, Netherlands: Kluwer Academic Publishers.

DME (Department of Minerals and Energy). 2001a. *Annual Report 2000 - 2001*. Pretoria: Government of South Africa.

DME. 2001b. *Strategy on Renewable Energy*. Draft. Pretoria: Government of South Africa.

DME. 1998. *White Paper on Energy Policy for South Africa*. Pretoria: Government of South Africa.

DTI (Department of Trade and Industry). 1996. *Growth, Employment and Redistribution: A Macro-Economic Strategy*. Pretoria: Government of South Africa.

Dubash, N. ed. 2002. *Power Politics: Equity and Environment in Electricity Reform*. Washington, DC: World Resources Institute.

EIA (Energy Information Administration). 2002a. *International Energy Outlook 2002*. Washington, DC: U.S. Energy Information Administration. Online at: http://www.eia.doe.gov/oiaf/ieo/index.html (June 27, 2002).

EIA. 2002b. *International Energy Annual 2000*. Washington, DC: U.S. Department of Energy. Online at: http://www.eia.doe.gov/emeu/iea (August 8, 2002).

EIA. 2001a. *International Energy Annual 1999*. Washington, DC: U.S. Department of Energy.

EIA. 2001b. *International Energy Outlook 2001*. Washington, DC: U.S. Department of Energy.

EIA. 1999. *International Energy Outlook 1999*. Washington, DC: U.S. Department of Energy.

Elzen, M.G. den, et al. 1999. *The Brazilian Proposal and Other Options for International Burden Sharing: An Evaluation of Methodological and Policy Aspects Using the FAIR model*. Report No. 728001011. Bilthoven, Netherlands: RIVM. Online at: http://www.unfccc.int/sessions/workshop/010528/rap728001011.pdf (June 27, 2002).

Elzen, M.G. den, and M. Schaeffer. 2002. "Responsibility for Past and Future Global Warming: Uncertainties in Attributing Anthropogenic Climate Change." *Climatic Change* 54: 29—73.

Embree, S. and K. Wilkinson. 1999. "Voluntary Commitments and the Kyoto Mechanisms: Exploring a 'Third Way' and Other Options."

ENDA-TM. 2001. *Climate Change: African Points of View*. Dakar: Environmental Development Action in the Third World.

Enting, I.G. and R.M. Law. 2002. "Characterising Historical Responsibility for the Greenhouse Effect." CSIRO Atmospheric Research Technical Paper no. 41. Online at: http://www.dar.csiro.au/publications/Enting_2002a.pdf

Epstein, J. and R. Gupta. 1990. *Controlling the Greenhouse Effect: Five Global Regimes Compared*. Washington, DC: Brookings Institution.

Eskom. 2000a. *Efficient Lighting Initiative*. Menlo Park.

Eskom. 2000b. *Environmental Report 2000: Towards Sustainability*. Sandton.

European Commission. 1996. *Poles 2.2*. Joule II Programme, Directorate General XII Science, Research and Development EUR 17358 EN.

Figueres, C., ed. 2002. *Establishing National Authorities for the CDM: A Guide for Developing Countries*. Winnepeg: International Institute for Sustainable Development.

Frankel, J. 1999. "Greenhouse Gas Emissions." *Policy Brief* #52. Washington, DC: Brookings Institution. Online at: http://www.brookings.org/dybdocroot/comm/PolicyBriefs/pb052/pb52.htm (June 27, 2002).

GCI (Global Commons Institute). 2002. "The Detailed Ideas and Algorithms Behind Contraction and Convergence." Online at: http://www.gci.org.uk/contconv/Ideas_behind_cc.html. Other useful materials on Contraction and Convergence are online at: http://www.gci.org.uk/contconv/cc.html and http://www.gci.org.uk/papers/papers.html (Papers and Reviews)

GCI. 1999. "Climate Change - A Global Problem, Contraction & Convergence - A Global Solution." Online at: http://www.gci.org.uk.

Girardin, L.O. 2000. "El Cambio Climático Global y la Distribución de los Costos de Mitigación de sus Eventuales Consecuencias Entre los Distintos Países." University of Buenos Aires. Faculty of Philosophy and Letters. Geographic Institute. Masters in Environmental and Territorial Policies. Master's Thesis. March.

Gobierno de la República Argentina. 1999. *Estudio sobre los Mecanismos de Flexibilización dentro del Contexto de la Convención Marco de las Naciones Unidas para el Cambio Climático y el Protocolo de Kioto*. Ministerio de Relaciones Exteriores Comercio Internacional y Culto, World Bank, Ministry of the Environment of the Republic of Canada. Final Report.

Government of Canada. 1999. *Environment Canada: What is Sustainable Development?* Online at: http://www.sdinfo.gc.ca/Eng/hist_e.cfm?intID=124 (June 27, 2002).

Government of Mexico. 2001. *Second National Communications to the United Nations Framework Convention on Climate Change*. D.F. México: Comité Intersecretarial sobre Cambio Climático.

Groenenberg, H., D. Phylipsen, and K. Blok. 2001. "Differentiating Commitments World Wide: Global Differentiation of GHG Emissions Reductions Based on the Triptych Approach Preliminary Assessment." *Energy Policy* 29: 1007—1030.

Grubb, M. 1995. "Seeking Fair Weather: Ethics and the International Debate on Climate Change." *International Affairs* 71(3): 463—495.

Grubb, M. 1989. *The Greenhouse Effect: Negotiating Targets*. London: The Royal Institute of International Affairs.

Grubb, M., C. Vrolijk, and D. Brack. 1999. *The Kyoto Protocol: A Guide and Assessment*. London: Royal Institute for International Affairs, Energy and Environmental Programme.

Gupta, J. 1997. *The Climate Change Convention and Developing Countries: From Conflict to Consensus?* Dordrecht, Netherlands: Kluwer Academic Publishers.

Gupta, S. and P. Bhandari. 1999. "An Effective Allocation Criterion for CO_2 Emissions." *Energy Policy* 27: 727—736.

Harris, P.G. 2000. "Defining International Distributive Justice: Environmental Considerations." *International Relations* 15(2): 51—66.

Helio International. 2000. *Sustainable Energy Watch: Methodology*. Paris.

Hendler, P. 2000. Data supplied by Paul Hendler, Housing Solutions, based on primary research with developers and various secondary sources. Johannesburg: Housing Solutions.

Höhne, N. and Harnisch, J. 2002. "Evaluating Indicators for the Relative Responsibility for Climate Change - Alternatives to the Brazilian Proposal and Global Warming Potentials," in *Non-CO_2 Greenhouse Gases*, Van Ham et al., eds. Rotterdam: Milpress.

Houghton, R.A. and J.L. Hackler. 2000. "Carbon Flux to the Atmosphere from Land-Use Changes," in *Trends: A Compendium of Data on Global Change*. Oak Ridge, Tennessee: Carbon Dioxide Information Analysis Center, Oak Ridge National Laboratory, U.S. Department of Energy. Online at: http://cdiac.esd.ornl.gov/trends/landuse/houghton/houghton.html (June 27, 2002).

Howells, M.I. 2000. *Baseline and Greenhouse Gas Mitigation Options for Bulk Energy Supply*. Cape Town: University of Cape Town, Energy Research Institute.

IEA (International Energy Agency). 2001. *Key World Energy Statistics from the IEA*. Paris.

IEA. 2000. CO_2 *Emissions from Fuel Combustion 1971-1998*. Paris.

IPCC (Intergovernmental Panel on Climate Change). 2001a. *Climate Change 2001: The Scientific Basis*. Summary for Policymakers. Cambridge: Cambridge University Press.

IPCC. 2001b. *Climate Change 2001: Impacts, Adaptation and Vulnerability*. Summary for Policymakers, J. McCarthy et al., eds. Cambridge: Cambridge University Press.

IPCC. 2001c. *Climate Change 2001: Mitigation*. Contribution of Working Group III to the Third Assessment Report of the IPCC, B. Metz and O. Davidson, eds. Cambridge: Cambridge University Press.

IPCC. 2001d. *Summary for Policymakers to Climate Change 2001*. Synthesis Report of the IPCC Third Assessment Report. Cambridge: Cambridge University Press.

IPCC. 2000a. *Special Report on Emissions Scenarios*. A special report of Working Group III of the IPCC, N. Nakicenovic and R. Swart, eds. Cambridge: Cambridge University Press.

IPCC. 2000b. *Special Report on Land Use, Land Use Change, and Forestry*. Cambridge: Cambridge University Press.

Irurah, D.K., ed. 2000. *Environmentally Sound Energy Efficient Low-Cost Housing for Healthier, Brighter and Wealthier Households, Municipalities and Nation*. Final report. Pretoria: Environmentally Sound Low-Cost Housing Task Team and the United States Agency for International Development.

KEEI (Korea Energy Economics Institute). 2000. *Study on Strategies to Address the UNFCCC and the Kyoto Protocol*. Seoul.

KEEI. 1990. *21st Century Energy Policy*. Seoul.

La Rovere, E.L. 2002 forthcoming. "Climate Change and Sustainable Development Strategies: a Brazilian Perspective." Paper commissioned by the OECD.

La Rovere, E.L. 1998. "Climate Change Convention: a Tool of Sustainable Development?" *Economies et Sociétés*, Série F, No. 36, 1/1998: 247—259.

Laitner, S. 2001. "Energy Efficiency Investments: A Dynamic Link between Environmental Quality and Positive Job Benefits for South Africa." Paper for the Domestic Use of Energy Conference, Cape Town, South Africa, April 10—12.

Levy, M.A., R.O. Keohane, and P.M. Haas. 1992. "Institutions for the Earth: Promoting International Environmental Protection." *Environment* 34(4).

Lutter, R. 2000. "Developing Countries' Greenhouse Emissions: Uncertainty and Implications for Participation in the Kyoto Protocol." *The Energy Journal* 21(4): 93—120.

Majola, B. 2002. "Highs and Lows of State's Delivery," *Mail and Guardian*. February 8–14: 10.

Marland, G., T.A. Boden, and R. J. Andres. 2000. "Global, Regional, and National CO_2 Emissions," in *Trends: A Compendium of Data on Global Change*. Oak Ridge, Tennessee: Carbon Dioxide Information Analysis Center, Oak Ridge National Laboratory, U.S. Department of Energy. Online at: http://cdiac.esd.ornl.gov/trends/emis/meth_reg.htm (June 27, 2002).

MCT (Ministério da Ciência e Tecnología). 1998. "Intervention of the Brazilian Delegation on Item 6 of the Provisional Agenda 'Voluntary Commitments.'" Government of Brazil.

Mehlwana, M.A. and N. Qase. 1999. *The Contours of Domesticity, Energy Consumption and Poverty: The Social Determinants of Energy Use in Low-Income Urban Households in Cape Town's Townships (1995-1997)*. Cape Town: Energy & Development Research Centre, University of Cape Town.

Menand, L., 2001. *The Metaphysical Club: A Story of Ideas in America*. New York: Ferrar, Strauss and Giroux.

Meyer, A. 2000. *Contraction and Convergence: The Global Solution to Climate Change*. Schumacher Briefing No. 5. Devon, UK: Green Books.

Michaelowa, A. and M. Dutschke. 2000. *Climate Policy and Development: Flexible Instruments and Developing Countries*. Cheltenham, UK: Edward Elgar.

Ministry of Science and Technology. 2000. "Notes on the Time-Dependent Relationship between Emissions of Greenhouse Gases and Climate Change." Government of Brazil.

Ministry of Science and Technology. 1999. "Proposed Elements of a Protocol to the UNFCCC Presented by Brazil in Response to the Berlin Mandate." Government of Brazil. Online at: http://www.mct.gov.br.gabin/clima.htm.

MIT (Massachusetts Institute of Technology). 1997. *Emissions Prediction and Policy Analysis Model*. Cambridge, Massachusetts: Center for Global Change Science.

Mlambo-Ngcuka, P. 2002. Minister of Minerals and Energy. *Parliamentary media briefing*. February 15. Cape Town: Ministry of Minerals and Energy.

Montreal Protocol on Substances that Deplete the Ozone Layer. 2000. Nairobi: UNEP Ozone Secretariat. Available online: http://www.unep.org/ozone/pdf/Montreal-Protocol2000.pdf (June 27, 2002).

Morita, T. and J. Robinson. 2001. "Greenhouse Gas Emission Mitigation Scenarios and Implications," in *Climate Change 2001: Mitigation*. Contribution of Working Group III to the Third Assessment Report of the Intergovernmental Panel on Climate Change, B. Metz et al., eds. Cambridge: Cambridge University Press.

Müller, B. 2001a. "Fair Compromise in a Morally Complex World." Paper presented at Pew Equity Conference, Washington, DC, April 17—18.

Müller, B. 2001b. "Varieties of Distributive Justice in Climate Change." *Climatic Change* 48(2–3).

Munasinghe, M. 2001. "Sustainable Development and Climate Change: Applying the Sustainomics Transdisciplinary Meta-Framework." *International Journal of Global Environmental Issues* 1(1): 13—55.

Mwandosya, M.J. 2000. *Survival Emissions: A Perspective from the South on Global Climate Change Negotiations*. Dar es Salaam: Dar es Salaam University Press and the Centre for Energy, Environment, Science and Technology.

NEPAD (New Partnership for Africa's Development). 2001. Online at: www.gov.za/issues/nepad.htm.

NER (National Electricity Regulator). 2000. *Electricity Supply Statistics for South Africa 2000*. Pretoria.

OECD/IEA (OECD Environment Directorate and the International Energy Agency). 2001. *An Initial View on Methodologies for Emission Baselines: Transport Case Study*. Information Paper. Paris.

Ott, H. and W. Sachs. 2000. *Ethical Aspects of Emissions Trading*. Germany: Wuppertal Institute.

Pearce, D.W., ed. 1986. *Macmillan Dictionary of Modern Economics*. Third Edition. London: Macmillan.

Philibert, C. 2002a. "Evolution of Mitigation Commitments: Certainty Versus Stringency." OECD/IEA Project for the Annex I Expert Group on the UNFCCC. Draft Paper, personal communication.

Philibert, C. 2002b. "Evolution of Mitigation Commitments: Fixed Targets Versus More Flexible Architectures." OECD/IEA Project for the Annex I Expert Group on the UNFCCC. Draft Paper, personal communication.

Philibert, C. and J. Pershing. 2001. "Considering the Options: Climate Targets for All Countries?" *Climate Policy* 1: 211—227.

Philibert, C. and J. Pershing. 2000. "Evolution of Mitigation Commitments." Draft paper, personal communication.

Pinguelli Rosa, L., S.K. Ribeiro, and M.S. Muylaert. 2001. "Comments on the Brazilian Proposal and Contributions for Annex I and Non-Annex I CO_2 Emissions from Fossil Fuels to Global Temperature Increase with Different Climate Responses." Unpublished paper.

Pizer, W. 1999. "Choosing Price or Quantity Controls for Greenhouse Gases." *Climate Issues Brief*. No. 17. Washington, DC: Resources for the Future.

Raskin, P. et al. 1998. *Bending the Curve: Toward Global Sustainability. A Report of the Global Scenario Group*. Boston: Stockholm Environment Institute.

Redefining Progress. 1999. *Fair and Low-Cost Climate Protection*. San Francisco.

Reid, W.V. and J. Goldemberg, eds. 1999. *Promoting Development While Limiting Greenhouse Gas Emissions: Trends and Baselines*. New York: UNDP and WRI.

Reiner, D. and H. Jacoby. 1997. *Annex I Differentiation Proposals: Implications for Welfare, Equity and Policy*. Boston: MIT.

Ringius, L., A. Torvanger, and A. Underdal. 2000. *Burden Differentiation: Fairness Principles and Proposals*. The joint CICERO-ECN project on sharing the burden of GHG reduction. Oslo: CICERO.

Rose, A. and B. Stevens. 1998. "A Dynamic Analysis of Fairness in Global Warming Policy: Kyoto, Buenos Aires and Beyond." *Journal of Applied Economics* 1(2): 329—362.

Rose, A., B. Stevens, and J. Edmonds. 1998. "International Equity and Differentiation in Global Warming Policy: An Application to Tradable Emission Permits." *Environmental & Resource Economics* 12(1): 25—51.

Roy, J. 2000. "The Rebound Effect: Some Empirical Evidence from India." *Energy Policy* 28(6–7): 433-438.

SA-CAN (South African Climate Action Network). 2002. *Sustainable Energy News by E-mail*. February.

Sachs, W. 1999. *Planet Dialectics: Explorations in Environment and Development*. London: Zed Books.

Sari, A.P. 1998. *On Equity and Developing Country Participation*. Berkeley: Energy and Resources Group, University of California, Berkeley.

Sijm, J.P.M. et al. 2000. *The Multi-Sector Convergence Approach of Burden Sharing: An Analysis of its Cost Implications*. Oslo: CICERO.

Simmonds, G. and Mammon, N. 1996. *Energy Services in Low-Income Urban South Africa: A Quantitative Assessment*. Cape Town: Energy & Development Research Centre, University of Cape Town.

Smith, K.R. and D.R. Ahuja. 1990. "Toward a Greenhouse Gas Equivalence Index: The Total Exposure Analogy." *Climatic Change* 17: 1–7.

Smith, K.R., J. Swisher, R. Kanter, and D.R. Ahuja. 1990. "Indices for a Greenhouse Gas Control Regime: Incorporating Both Efficiency and Equity Goals." Prepared for the Environmental Policy and Research Division of the World Bank.

Sokona, Y., S. Humphreys, and J.P. Thomas. 1999. "Development: A Centrepiece of the Kyoto Protocol, an African Perspective." Fourth Conference of the Parties for the UNFCCC, Buenos Aires, 8 November.

South African Reserve Bank. 2001. *Annual Report* and *Quarterly Bulletin*. Pretoria. Online at: http://www.resbank.co.za/.

Spalding-Fecher, R. 2001. *Energy and Sustainability in South Africa: the 2001 Sustainable Energy Watch Report*. Cape Town: Energy & Development Research Centre, University of Cape Town.

Spalding-Fecher, R, A. Clark, M. Davis, and G. Simmonds. 2002, submitted. "The Economics of Energy Efficiency for the Poor—a South African Case Study." *Energy: The International Journal*.

Stavins, R.N. 2001. *Lessons from the American Experiment with Market-Based Environmental Policies*. Harvard Kennedy School Working Papers, HRWP01-032. Cambridge, MA: Harvard University.

TERI (Tata Energy Research Institute). 1997. *Long Term Carbon Emission Targets Aiming Towards Convergence*. New Delhi.

Thorne, S. and E. La Rovere. 1999. "Criteria and Indicators for the Appraisal of Clean Development (CDM) Projects." Paper presented at the Fifth Conference of the Parties to the Framework Convention on Climate Change, Helio International, Bonn, November 26.

Tietenberg, T. 1985. *Emissions Trading: An Exercise in Reforming Pollution Policy*. Washington, DC: Resources for the Future.

Toman, M. and M. Cazorla. 2000. *International Equity & Climate Change Policy*. Washington, DC: Resources for the Future.

Torvanger, A. and O. Godal. 1999. *A Survey of Differentiation Methods for National Greenhouse Gas Reduction Targets*. Oslo: CICERO.

Toth, F. and M. Mwandosya. 2001. "Decision-Making Frameworks," in *Climate Change 2001: Mitigation*. Contribution of Working Group III to the Third Assessment Report of the Intergovernmental Panel on Climate Change, B. Metz et al., eds. Cambridge: Cambridge University Press.

Tynkkynen, O. 2000. *Framework for Fair Shares - Equity in Burden Sharing*. Washington, DC: Friends of the Earth.

UCS (Union of Concerned Scientists). 2001. *Clean Energy Blueprint*. Washington, DC.

UCT (University of Cape Town). 2002. *Options for a Basic Electricity Support Tariff: Analysis, Issues and Recommendations*. For the Department of Minerals and Energy and Eskom. Cape Town.

UNFCCC (United Nations Framework Convention on Climate Change). 2002. "Report of the Conference of the Parties on its Seventh Session, Held at Marrakesh from 29 October to 10 November 2001." Decision 15/CP.7. Document numbers: FCCC/CP/2001/13/Add.1 to Add.4.

UNFCCC. 2001. *Scientific and Methodological Aspects of the Proposal by Brazil*. Progress report on the review of the scientific and methodological aspects of the proposal by Brazil (Note by the secretariat). Document number: FCCC/SBSTA/2001/INF.2, July 10. Online at: http://www.unfccc.int/resource/docs/2001/sbsta/inf02.pdf (June 27, 2002).

UNFCCC. 2000a. "In-Depth Review of the Second National Communication of Bulgaria." United Nations Framework Convention on Climate Change. Online at: http://www.unfccc.int.

UNFCCC. 2000b. "Submission of India on Article 12 of the Kyoto Protocol." Document number: FCCC/SB/2000/Misc.1. Online at: http://unfccc.int.

UNFCCC, 2000c. *Tracing the Origins of the Kyoto Protocol: An Article-by-Article Textual History*. Technical Paper prepared by Joanna Depledge. Document FCCC/TP/2000/2.

UNFCCC. 1999. "Review of the Implementation of Commitments and of Other Provisions of the Convention." Document number: FCCC/CP/1999/7.

UNFCCC. 1998. "Group of 77 and China Position Paper on Preparations for COP4 (Proposed Provisional Agenda)." Submission by the Group of 77 and China. Document number: FCCC/SBI/1998/MISC.3.

UNFCCC. 1997a. *Kyoto Protocol to the United Nations Framework Convention on Climate Change*. Bonn: UNFCCC Secretariat. Online at: http://unfccc.int/resource/convkp.html (June 27, 2002).

UNFCCC. 1997b. "Paper no. 1: Brazil: Proposed Elements of a Protocol to the United Nations Framework Convention on Climate Change." Presented by Brazil in Response to the Berlin Mandate, UNFCCC/AGBM/1997/MISC.1/Add.3 GE.97-, Bonn. Online at: http://www.unfccc.de (and in Portuguese: http://www.mct.gov.br/clima/).

UNFCCC. 1992. New York: United Nations. Full treaty text available online at: http://unfccc.int/resource/conv/index.html (June 27, 2002).

United Nations (UN). 2000. "Millennium Declaration: Resolution Adopted by the General Assembly." New York.

United Nations. 1998. "Climate Change and Desertification." Statement by Minister Enio Cordeiro in New York, United Nations General Assembly, Fifty-Third Session, Second Committee, October 22.

US EPA (United States Environmental Protection Agency), Office of Policy Analysis. 1985. *Costs and Benefits of Reducing Lead in Gasoline, Final Regulatory Impact Analysis*. Washington, DC.

van der Merwe, M.R. and R.J. Scholes. 1998. *South African Greenhouse Gas Emissions Inventory for the Years 1990 and 1994*. Pretoria: National Committee on Climate Change.

Victor, D. 2001. *The Collapse of the Kyoto Protocol and the Struggle to Slow Global Warming*. Council on Foreign Relations. Princeton, NJ: Princeton University Press.

Victor, D.G. 1999. "The Regulation of Greenhouse Gases: Does Fairness Matter?" in *Fair Weather? Equity Concerns in Climate Change*, F.L. Toth, ed. London: Earthscan.

Waldron, A. 2002. "China's Economic Façade," *Washington Post* (21 March, p. A35).

WCD (World Commission on Dams Secretariat). 2000. *Dams and Global Climate Change*. Thematic Review II.2 prepared as an input to the WCD. Cape Town.

Werksman, J. 1999. "Procedural and Institutional Aspects of the Emerging Climate Regime: Do Improvised Procedures Lead to Impoverished Rules?" Paper presented at Concluding Workshop for the Project to Enhance Capacity under the Framework Convention on Climate Change and the Kyoto Protocol, London, March 17—18.

White House. 2002. *Fact Sheet: President Bush Announces Clear Skies & Global Climate Change*. Washington, DC. Online at: http://www.whitehouse.gov/news/releases/2002/02/20020214.html (June 27, 2002).

Wiener, J.B. 1999. "Global Environmental Regulation: Instrument Choice in a Legal Context." *Yale Law Journal* 108: 677.

Winkler, H. and J. Mavhungu. 2001. *Green Power, Public Benefits and Electricity Industry Restructuring*. Cape Town: University of Cape Town, Energy & Development Research Centre.

Winkler, H., R. Spalding-Fecher, and L. Tyani. 2001. *What Could Potential Carbon Emissions Allocation Schemes and Targets Mean for South Africa?* Cape Town: University of Cape Town, Energy & Development Research Centre.

Winkler, H., R. Spalding-Fecher, L. Tyani, and K. Matibe. 2000. *Cost-Benefit Analysis of Energy-Efficiency in Low-Cost Housing*. Cape Town: University of Cape Town, Energy & Development Research Centre.

Winkler, H. and S. Thorne. 2002. "Baselines for Suppressed Demand: Climate Change Contributions to Poverty Alleviation Through CDM Projects." Annual Conference of the Forum for Economics and Environment, Cape Town, February 11–12.

World Bank. 2002. "World Development Indicators." Online at http://www.worldbank.org/data/wdi2000/. Washington, DC.

World Bank. 2000. *World Development Report 2000/2001*. Oxford: Oxford University Press.

WRI (World Resources Institute). 2002. "Analysis of Bush Administration Greenhouse Gas Target." Online at: http://www.wri.org/pdf/analysis_bush.pdf (June 27, 2002).

Ybema, J. et al. 2000. *Burden Differentiation: GHG Emissions, Undercurrents and Mitigation Costs*. Oslo, Norway: CICERO.

Zhang, Z.X. 2001. "Meeting the Kyoto Targets: The Importance of Developing Country Participation." Nota di Lavoro 30-2001. Milan: Fondazione Eni Enrico Mattei.

Zhang, Z.X. 2000. "Estimating the Size of the Potential Market for the Kyoto Flexibility Mechanisms." *Weltwirtschaftliches Archiv - Review of World Economics* 136(3): 491—521.

Zhou, P. 2001. "North-South Dialogue." *Tiempo: Global Warming and the Third World* 40/41: 1—7.

Glossary and Abbreviations

Annex I Parties. The industrialized and transitioning countries listed in this Annex to the Climate Convention. These countries accepted emission targets for the period 2008 to 2012 in Annex B of the Kyoto Protocol.

Annex II Parties. The wealthy countries listed in this Annex to the Climate Convention that have a special obligation to help developing countries with financial and technological resources. They include the 24 original members of the Organization for Economic Cooperation and Development (OECD) plus the European Union.

Annex B. An Annex to the Kyoto Protocol that lists agreed emission targets taken by the industrialized and transitioning countries for the so-called first commitment period, which runs from 2008 to 2012.

AOSIS. Alliance of Small Island States. An ad hoc coalition of 42 low-lying and island countries that are particularly vulnerable to sea-level rise and share common positions on climate change.

BAU. Business as usual. A scenario that represents the most plausible projection of the future. BAU embodies the notion of what would happen, hypothetically, if climate-friendly actions were not taken.

Berlin Mandate. An agreement reached in 1995 in Berlin, Germany, at the first Conference of the Parties to the Climate Convention, in which the industrialized countries first agreed to take on targets and timetables for quantified reductions and limitations on greenhouse gas emissions.

Capacity Building. Increasing skilled personnel and technical and institutional abilities.

CDF. Clean Development Fund. An element of the original Brazilian Proposal that was adapted to become the Clean Development Mechanism of the Kyoto Protocol.

CDM. Clean Development Mechanism. A project-based emissions trading system under the Kyoto Protocol that allows industrialized countries to use emission reduction credits from projects in developing countries that both reduce greenhouse gas emissions and promote sustainable development.

Climate Convention. See UNFCCC.

COP. Conference of the Parties to the Climate Convention. The supreme body of the Convention. It currently meets once a year to review the Convention's progress. The word "conference" is not used here in the sense of "meeting" but rather of "association," which explains the seemingly redundant expression "fourth session of the Conference of the Parties."

COP/MOP. Conference of the Parties serving as the Meeting of the Parties to the Protocol. The Kyoto Protocol's supreme body, which will serve as the Protocol's meeting of the Parties. The sessions of the COP and the COP/MOP will be held during the same period. This will improve cost-effectiveness and coordination with the Convention.

CO_2. Carbon dioxide, a naturally occurring gas. It is also a by-product of burning fossil fuels and biomass and other industrial processes as well as land use changes. CO_2 is the principal anthropogenic greenhouse gas affecting the Earth's temperature.

EIT. Economy in transition. EITs typically include the countries of Central and Eastern Europe (e.g., Poland), the former Soviet Union (e.g., Russia), and Central Asian Republics (e.g., Kazakhstan).

EU. European Union. Includes 15 member states.

GDP. Gross domestic product. The total value of goods and services produced by an economy.

GEF. Global Environment Facility. The designated financial mechanism for international agreements on biodiversity, climate change (i.e., the UNFCCC), and persistent organic pollutants. Established in 1991, the GEF helps developing countries fund projects and programs that protect the global environment.

Global Warming Potential (GWP). An index that allows for comparison of the various greenhouse gases. It is the radiative forcing that results from the addition of 1 kilogram of a gas to the atmosphere compared to an equal mass of carbon dioxide. Over 100 years, methane has a GWP of 21 and nitrous oxide of 310.

Greenhouse Effect. The effect produced as greenhouse gases allow incoming solar radiation to pass through the Earth's atmosphere but prevent most of the outgoing long-wave infrared radiation from the surface and lower atmosphere from escaping into outer space. This envelope of heat-trapping gases keeps the Earth about 30° C warmer than if these gases did not exist.

GHG. Greenhouse gas. Any gas that absorbs and re-emits infrared radiation into the atmosphere. The main greenhouse gases include water vapor (H_2O), carbon dioxide (CO_2), methane (CH_4), and nitrous oxide (N_2O).

G-77. Group of 77. Founded in 1967 under the auspices of the United Nations Conference for Trade and Development (UNCTAD); seeks to harmonize the negotiating positions of its 133 developing-country members.

Hot Air. A term developed by the nongovernmental community to describe the gap between a country's emissions target agreed to under the Kyoto Protocol and its actual emissions, when its emissions have been reduced due to reasons unrelated to climate change mitigation. The changes in the economies of reunified Germany, Russia, and other economies in transition, for example, have resulted in emission trajectories that are below their Kyoto targets.

IMF. International Monetary Fund. An international organization of 184 member countries established to promote international monetary cooperation and foster economic growth.

IPCC. Intergovernmental Panel on Climate Change. An organization established in 1988 by the World Meteorological Organization and the United Nations Environment Programme. It conducts rigorous surveys of the worldwide technical and scientific literature and publishes assessment reports widely recognized as the most credible existing sources on climate change.

JI. Joint implementation. The mechanism established by the Kyoto Protocol whereby an Annex I country can receive "emissions reductions units" when it helps to finance projects that reduce net emissions in another Annex I country.

Kyoto Protocol. An international agreement adopted by all Parties to the Climate Convention in Kyoto, Japan, in December 1997.

LDC. Least developed country. A category of countries (currently 49) deemed by the United Nations to be structurally handicapped in their development process, facing more than other developing countries the risk of failing to come out of poverty as a result of these handicaps, and in need of the highest degree of consideration from the international community in support of their development efforts.

Marrakesh Accords. A set of detailed rules for implementation of the Kyoto Protocol, adopted in 2001 at the seventh Conference of the Parties in Marrakesh, Morocco.

OECD. Organization for Economic Cooperation and Development. An international organization consisting of the major industrialized countries. The OECD includes Mexico and the Republic of Korea, which are non-Annex I countries under the Kyoto Protocol.

Party. A state (or regional economic integration organization, such as the European Union) that agrees to be bound by a treaty and for which the treaty has entered into force.

PAMs. Policies and measures. The promotion of renewable energy, energy efficiency, forest conservation, or other actions for the reduction or limitation of greenhouse gases or for sustainable development.

ppmv. Parts per million by volume. A unit of concentration for a particular substance (e.g., CO_2).

PPP. Purchasing power parity. An international dollar "currency" for GDP that has the same purchasing power over local GDP as a U.S. dollar has in the United States.

SBI. Subsidiary Body for Implementation. An official body of the Climate Convention, open to all Parties, that makes recommendations on policy and implementation issues to the Conference of the Parties and, if requested, other bodies.

SBSTA. Subsidiary Body for Scientific and Technological Advice. An official body of the Climate Convention, open to all Parties, that serves as the link between the information and assessment provided by expert sources (such as the Intergovernmental Panel on Climate Change) on the one hand, and the policy-oriented needs of the Conference of the Parties on the other.

S-CDM. Sector-Based Clean Development Mechanism. An approach to expanding the Clean Development Mechanism to encompass entire sectors, geographic regions, and combinations of sectors and regions. See Chapter 4.

SD-PAMs. Sustainable Development Policies and Measures. An approach to climate protection that builds on sustainable development priorities. See Chapter 3.

Sequestration. Absorption of carbon dioxide from the atmosphere through the process of photosynthesis.

UNFCCC. United Nations Framework Convention on Climate Change (Climate Convention, or Convention). A treaty signed at the 1992 Earth Summit in Rio de Janeiro by more than 150 countries.

Website Sources: UNFCCC (http://unfccc.int/siteinfo/glossary.html); UNCTAD (http://www.unctad.org/ldcs); World Bank (www.worldbank.org/data/); Center for Sustainable Development in the Americas (http://www.csdanet.org/English/publications/glossary.htm).

About the Authors

Mr. Malik Amin Aslam is the executive director for ENVORK, a Pakistan-based research and development organization focusing on climate change-related issues.

Mr. Kevin A. Baumert is a senior associate in the Climate, Energy and Pollution Program at the World Resources Institute.

Dr. Odile Blanchard is an associate professor of economics at the University of Grenoble, France, and a researcher at Institut d'Economie et de Politique de l'Energie (IEPE). With the support of Institut Francais de l'Energie, Dr. Blanchard was a Visiting Fellow at WRI during 2001-2002.

Dr. Daniel Bouille is vice president and senior researcher at the Institute for Energy Economics at the Bariloche Foundation in Argentina. Dr. Bouille participated in defining the voluntary target for Argentina.

Dr. Ogunlade Davidson is professor of engineering and director of Energy and Development Research Centre at the University of Cape Town, and also the co-chair of Working Group III (Mitigation) of the Intergovernmental Panel on Climate Change.

Dr. Joanna Depledge is honorary research fellow at the Department of Geography, University College London. She is a former staff member of the UNFCCC Secretariat.

Ms. Christiana Figueres is the executive director of the Center for Sustainable Development in the Americas (CSDA).

Dr. Osvaldo Girardin is director of the Environmental and Development Program at the Bariloche Foundation in Argentina. Dr. Girardin participated in defining the voluntary target for Argentina.

Dr. Nancy Kete is a senior fellow at WRI and co-director of EMBARQ—the WRI Center for Transport and the Environment.

Dr. Yong-Gun Kim is a research fellow in the Global Environment Research Center at the Korea Environment Institute.

Ms. Silvia Llosa was formerly an associate in the Climate, Energy and Pollution Program at the World Resources Institute.

Dr. Emilio La Rovere is professor of the Energy Planning Program and coordinator of the Center for Integrated Studies on Climate Change and the Environment at COPPE/UFRJ.

Mr. Stanford Mwakasonda is a senior researcher at the Energy and Development Research Centre at the University of Cape Town, focusing on energy and climate change.

Mr. James F. Perkaus is a fellow in the Climate, Energy and Pollution Program at WRI and a Ph.D. candidate in international relations at Tufts University's Fletcher School of Law and Diplomacy.

Mr. JoséLuis Samaniego is director of the Center for Sustainable Transport of México City and Commissioner for International Affairs in the Interdisciplinary Center for Biodiversity and Environment (CeIBA).

Mr. Randall Spalding-Fecher is a senior researcher at Energy and Development Research Centre at the University of Cape Town, focusing on energy, efficiency and environmental policy issues.

Ms. Laura Valente de Macedo is the Climate Change programme coordinator at Instituto Pró-Sustentabilidade (IPSUS) in São Paulo and the Latin America and the Caribbean Coordinator of the Cities for Climate Protection program at ICLEI in Rio de Janeiro. She is the former coordinator of the Brazilian Climate Change Forum.

Mr. Harald Winkler is a senior researcher at the Energy and Development Research Centre at the University of Cape Town, focusing on climate change (economics, mitigation, and policy) and environmental economics.

About WRI

The World Resources Institute is an environmental think tank that goes beyond research to create practical ways to protect the Earth and improve people's lives. Our mission is to move human society to live in ways that protect Earth's environment for current and future generations.

Our program meets global challenges by using knowledge to catalyze public and private action:

- *To reverse damage to ecosystems.* We protect the capacity of ecosystems to sustain life and prosperity.
- *To expand participation in environmental decisions.* We collaborate with partners worldwide to increase people's access to information and influence over decisions about natural resources.
- *To avert dangerous climate change.* We promote public and private action to ensure a safe climate and sound world economy.
- *To increase prosperity while improving the environment.* We challenge the private sector to grow by improving environmental and community well-being.

In all of our policy research and work with institutions, WRI builds bridges between ideas and action, meshing the insights of scientific research, economic analysis, and practical experience with the need for open and participatory decision-making.